T0214221

Lecture Notes in Computer Science 14528

Founding Editors

Gerhard Goos
Juris Hartmanis

Editorial Board Members

Elisa Bertino, *Purdue University, West Lafayette, IN, USA*
Wen Gao, *Peking University, Beijing, China*
Bernhard Steffen ⓘ, *TU Dortmund University, Dortmund, Germany*
Moti Yung ⓘ, *Columbia University, New York, NY, USA*

The series Lecture Notes in Computer Science (LNCS), including its subseries Lecture Notes in Artificial Intelligence (LNAI) and Lecture Notes in Bioinformatics (LNBI), has established itself as a medium for the publication of new developments in computer science and information technology research, teaching, and education.

LNCS enjoys close cooperation with the computer science R & D community, the series counts many renowned academics among its volume editors and paper authors, and collaborates with prestigious societies. Its mission is to serve this international community by providing an invaluable service, mainly focused on the publication of conference and workshop proceedings and postproceedings. LNCS commenced publication in 1973.

Michael Hartisch · Chu-Hsuan Hsueh ·
Jonathan Schaeffer

Editors

Advances in Computer Games

18th International Conference, ACG 2023
Virtual Event, November 28–30, 2023
Revised Selected Papers

 Springer

Editors
Michael Hartisch ⓘ
University of Siegen
Siegen, Germany

Jonathan Schaeffer
University of Alberta
Edmonton, AB, Canada

Chu-Hsuan Hsueh ⓘ
Japan Advanced Institute of Science
and Technology
Nomi, Ishikawa, Japan

ISSN 0302-9743   ISSN 1611-3349 (electronic)
Lecture Notes in Computer Science
ISBN 978-3-031-54967-0   ISBN 978-3-031-54968-7 (eBook)
https://doi.org/10.1007/978-3-031-54968-7

© The Editor(s) (if applicable) and The Author(s), under exclusive license
to Springer Nature Switzerland AG 2024

This work is subject to copyright. All rights are reserved by the Publisher, whether the whole or part of the material is concerned, specifically the rights of translation, reprinting, reuse of illustrations, recitation, broadcasting, reproduction on microfilms or in any other physical way, and transmission or information storage and retrieval, electronic adaptation, computer software, or by similar or dissimilar methodology now known or hereafter developed.
The use of general descriptive names, registered names, trademarks, service marks, etc. in this publication does not imply, even in the absence of a specific statement, that such names are exempt from the relevant protective laws and regulations and therefore free for general use.
The publisher, the authors, and the editors are safe to assume that the advice and information in this book are believed to be true and accurate at the date of publication. Neither the publisher nor the authors or the editors give a warranty, expressed or implied, with respect to the material contained herein or for any errors or omissions that may have been made. The publisher remains neutral with regard to jurisdictional claims in published maps and institutional affiliations.

This Springer imprint is published by the registered company Springer Nature Switzerland AG
The registered company address is: Gewerbestrasse 11, 6330 Cham, Switzerland

Paper in this product is recyclable.

Preface

This volume collects the papers presented at the 18th *Advances in Computer Games* conference (ACG 2023), which took place during 28–30 November 2023. The conference was held under the auspices of the International Computer Games Association, ICGA[1]. It was conducted online and coordinated between the University of Siegen (Germany), Japan Advanced Institute of Science and Technology (Japan), Maastricht University (the Netherlands), the University of Alberta (Canada), who provided the Zoom stream on which the conference was conducted, and IBM Research (Japan).

The biennial *Advances in Computer Games* conference series is a major international forum for researchers and developers interested in all aspects of artificial intelligence and computer game playing. Earlier conferences took place in London (1975), Edinburgh (1978), London (1981, 1984), Noordwijkerhout (1987), London (1990), Maastricht (1993, 1996), Paderborn (1999), Graz (2003), Taipei (2005), Pamplona (2009), Tilburg (2011), Leiden (2015, 2017), Macao (2019) and online (2021). For the past 20 years, the conference has been held every second year, alternating with the *Computers and Games* conference.

A total of 29 papers were submitted to ACG 2023, with 2 being desk-rejected and 14 accepted for presentation. In the single-blind review process, nearly every submission underwent at least three reviews, except for four submissions with two reviews available. The ACG 2023 program consisted of four keynote talks and five regular paper sessions, as listed below. All presentation videos are available online[2].

Session 1: Chess and Its Variants

The opening session, chaired by Todd Neller, focussed on game-playing programs in chess-related games. These included "Making Superhuman AI More Human in Chess" by Daniel Barrish, Steve Kroon and Brink van der Merwe, "Merging Neural Networks with Traditional Evaluations in Crazyhouse" by Anei Makovec, Johanna Pirker and Matej Guid, and "Stockfish or Leela Chess Zero? A Comparison Against Endgame Tablebases" by Quazi Asif Sadmine, Asmaul Husna and Martin Müller.

Keynote: Martin Müller *Solution Methods for Two Player Games*

The first keynote speaker was Martin Müller, the DeepMind Chair in Artificial Intelligence at the University of Alberta. He was introduced by Jonathan Schaeffer. Martin has spent thirty years advancing the fields of algorithms, artificial intelligence, and combinatorial game theory. His dedication to advancing heuristic search techniques and his profound impact on the theoretical and practical aspects of AI research are evident in numerous publications and successful applications. In his talk, he provided a comprehensive overview of the current status of solution methods in gaming. He covered games

---

that have been solved in the past 21 years, along with ongoing challenges that researchers are currently addressing as well as challenges that will be interesting to tackle in the future.

Session 2: Solving Games

This session, chaired by Ryan Hayward, presented new methods for game solving or new results of game analyses. The papers presented were "Solving NoGo on Small Rectangular Boards" by Haoyu Du, Ting Han Wei and Martin Müller and "Optimal Play of the Great Rolled Ones Game" by Todd W. Neller, Quan H. Nguyen, Phong T. Pham, Linh T. Phan and Clifton G. M. Presser.

Keynote: Frank Lantz *Games, Computability, AI, and Aesthetics*

The second keynote speech was given by Frank Lantz (NYU Game Center), introduced by Cameron Browne. Frank is a visionary game designer and founding chair of the New York University Game Center. With over 20 years of experience, he has not only shaped the landscape of game design but also co-founded influential game development entities and pioneered large-scale, real-world games. In his talk, he endeavored to redefine our perception of "aesthetics", urging us to view it as a broad category encompassing all games within the realm of human activity. He challenged us to recognize games not merely as entertainment but as an art form—an overarching aesthetic experience that transcends traditional boundaries.

Keynote: Hiroyuki Iida *Using Games to Study Law of Motions in Mind*

Ting Han Wei introduced the third keynote speaker, Hiroyuki Iida, the Vice-President of the Japan Advanced Institute of Science and Technology. Hiroyuki's extensive expertise lies in heuristic search, artificial intelligence, game-refinement theory and entertainment technology. Over the years, he has significantly contributed to the development of AI, particularly in the realm of game tree search and decision-making systems. He is well known for his contribution to the game of shogi. In his talk, he explored the concept of gravity in mind, shaping the balance between objectivity and subjectivity in gameplay. He suggested that when a game's outcome becomes predictable, a timely conclusion is crucial to prevent monotony, emphasizing the losing player's role in maximizing the game's artistic value. He argued that this concept can be utilized to provide and analyze exciting game experiences, ensuring that comfort and discomfort are in harmony.

Session 3: Board Games and Card Games

Kazuki Yoshizoe chaired this session on various investigations based on classical board games and card games. This session included the papers "MCTS with Dynamic Depth Minimax" by James Ji and Michael Thielscher, "Can We Infer Move Sequences in Go from Stone Arrangements?" by Chu-Hsuan Hsueh and Kokolo Ikeda, and "Quantifying Feature Importance of Games and Strategies via Shapley Values" by Satoru Fujii.

Session 4: Player Investigation

This session, chaired by Reijer Grimbergen, investigated player experience in video games. The papers included "The Impact of Wind Simulation on Perceived Realism of Players" by Zeynep Burcu Kaya Alpan and Şenol Pişkin and "Hades Again and Again: A Study on Frustration Tolerance, Physiology and Player Experience" by Maj Frost Jensen, Laurits Dixen and Paolo Burelli.

Keynote: Tristan Cazenave *Bootstrapping Artificial Intelligence*

The fourth and final keynote speaker was Tristan Cazenave (LAMSADE Université Paris Dauphine PSL CNRS), introduced by Jaap van den Herik. Tristan is a prominent figure in the field of artificial intelligence, particularly known for his work on computer games, artificial intelligence in games, and game-playing programs. He has made significant contributions to areas such as computer chess, computer Go, and other strategy games. He has been actively involved in research towards optimizing Monte Carlo searches for specific problem domains, accelerating AlphaZero-type deep reinforcement learning, and applying a fusion of Monte Carlo search and deep learning to diverse optimization challenges. In his talk, Tristan explored the concept of "Bootstrapping Artificial Intelligence", emphasizing the synergy between search and learning to enhance AI capabilities. He discussed diverse applications, from improving game-playing programs using introspective learning to algorithm discovery through the combination of Monte Carlo tree search and deep reinforcement learning.

Session 5: Math, Games, and Puzzles

Bruno Bouzy chaired Session 5, which collected research work on games involving math, analyses of games based on math, and new content generation for puzzles. The papers included "Analysis of a Collatz Game and Other Variants of the $3n + 1$ Problem" by Ingo Althöfer, Michael Hartisch and Thomas Zipproth, "Implicit QBF Encodings for Positional Games" by Irfansha Shaik, Valentin Mayer-Eichberger, Jaco van de Pol and Abdallah Saffidine, "The Mathematical Game" by Marc Pierre, Quentin Cohen-Solal and Tristan Cazenave, and "Slitherlink Art" by Cameron Browne.

Acknowledgements. The organization of ACG 2023 was made possible by the invaluable contributions of the authors, reviewers, session chairs and keynote speakers. Special thanks are due to Cameron Browne and Akihiro Kishimoto for their invaluable advice during the conference preparations. We thank the University of Alberta for providing online technology, supported by Danny Whittaker. Finally, we thank everyone who registered for ACG. Your attendance was critical to the success of this conference.

November 2023

Michael Hartisch
Chu-Hsuan Hsueh
Jonathan Schaeffer

Organization

Organizing Committee

Michael Hartisch	University of Siegen, Germany
Chu-Hsuan Hsueh	Japan Advanced Institute of Science and Technology, Japan
Jonathan Schaeffer	University of Alberta, Canada

Advisory Committee

Cameron Browne	Maastricht University, The Netherlands
Akihiro Kishimoto	IBM Research, Japan

ICGA Executive

Tristan Cazenave	LAMSADE Université Paris Dauphine PSL CNRS, France
Hiroyuki Iida	Japan Advanced Institute of Science and Technology, Japan
Mark Lefler	Independent, USA
David Levy	Independent, UK
Jonathan Schaeffer	University of Alberta, Canada
Jaap van den Herik	University of Leiden, The Netherlands
Mark Winands	Maastricht University, The Netherlands
I-Chen Wu	National Yang Ming Chiao Tung University, Taiwan

Program Committee

Yngvi Björnsson	Reykjavik University, Iceland
Bruno Bouzy	Paris Descartes University, France
Cameron Browne	Maastricht University, The Netherlands
Tristan Cazenave	LAMSADE Université Paris Dauphine PSL CNRS, France
Lung-Pin Chen	Tunghai University, Taiwan

Reijer Grimbergen	Tokyo University of Technology, Japan
Michael Hartisch	University of Siegen, Germany
Ryan Hayward	University of Alberta, Canada
Chu-Hsuan Hsueh	Japan Advanced Institute of Science and Technology, Japan
Hiroyuki Iida	Japan Advanced Institute of Science and Technology, Japan
Nicolas Jouandeau	Paris 8 University, France
Tomoyuki Kaneko	University of Tokyo, Japan
Akihiro Kishimoto	IBM Research, Japan
Jakub Kowalski	University of Wroclaw, Poland
Sylvain Lagrue	Université de Technologie de Compiégne, France
Martin Müller	University of Alberta, Canada
Todd Neller	Gettysburg College, USA
Mark Nelson	American University, USA
Diego Perez Liebana	Queen Mary University of London, UK
Éric Piette	Université catholique de Louvain, Belgium
Jonathan Schaeffer	University of Alberta, Canada
Matthew Stephenson	Flinders University, Australia
Ruck Thawonmas	Ritsumeikan University, Japan
Michael Thielscher	University of New South Wales, Australia
Jonathan Vis	University of Leiden, The Netherlands
Ting Han Wei	University of Alberta, Canada
Mark Winands	Maastricht University, The Netherlands
I-Chen Wu	National Yang Ming Chiao Tung University, Taiwan
Ti-Rong Wu	Academia Sinica, Taiwan
Kazuki Yoshizoe	Kyushu University, Japan

Contents

Math, Games, and Puzzles

Chess and Its Variants

Making Superhuman AI More Human in Chess

Daniel Barrish[1]([⊠]) [iD], Steve Kroon[1,2] [iD], and Brink van der Merwe[1,2] [iD]

[1] Computer Science Division, Stellenbosch University,
Stellenbosch, South Africa
`daniel.barrish@gmail.com`, {`kroon,abvdm`}`@sun.ac.za`
[2] National Institute for Theoretical and Computational Sciences,
Stellenbosch, South Africa

Abstract. Computer chess research has traditionally focused on creating the strongest possible chess engine. Recently, however, attempts have been made to create engines that mimic the playing strength and style of human players. Our research proposes enhancements of models developed in this vein that more accurately imitate master-level players, as well as improve the prediction accuracy of existing models on weaker players. Our proposed enhancements are simple to apply by post-processing the output of existing chess engines. The performance of our enhancements was evaluated and compared using two metrics, prediction accuracy and average centipawn loss. We found that using an ensemble model over search depths maximised prediction accuracy, while an evaluation window filtering approach was preferable with respect to average centipawn loss.

Keywords: Artificial intelligence · Chess · Action prediction

1 Introduction

Chess has famously been referred to as the "drosophila of artificial intelligence" for decades [4]. In other words, it is a testing ground well-suited to experimentation with new ideas and concepts in artificial intelligence, before applying very similar ideas and concepts to more complex real-world use cases. Computer chess research has primarily focused on creating the best chess engine possible, however recently interest has grown in developing models that mimic humans.

These efforts to develop a human-like engine have been spearheaded by the Maia chess team[1]. The Maia chess engine leverages existing work on the open-source engine Leela Chess Zero (Lc0)[2], which is based on DeepMind's AlphaZero [9]. In 2017, AlphaZero demonstrated the potential of using neural networks combined with reinforcement learning in chess. While AlphaZero and Lc0 are trained via self-play, Maia is trained on a database of human games

---

[1] https://maiachess.com.
[2] https://lczero.org.

© The Author(s), under exclusive license to Springer Nature Switzerland AG 2024
M. Hartisch et al. (Eds.): ACG 2023, LNCS 14528, pp. 3–14, 2024.
https://doi.org/10.1007/978-3-031-54968-7_1

instead. Although Maia achieved considerable success, it fell short in a number of areas—including the inability to accurately imitate strong human players, such as Grandmasters [5].

Our objectives are to overcome this shortcoming in particular, and improve upon the accuracy of Maia's move predictions in general. Since the Maia engines disable search by design, the main avenues explored in this research include the incorporation of search up to a fixed depth[3], a search depth ensemble method, and move selection based on an evaluation window.

2   Background

Since the dawn of the computer age, much effort has gone into creating the strongest chess engine possible. This work culminated in the famous 1997 'Man vs Machine' match, where IBM's Deep Blue beat the reigning World Champion, Garry Kasparov. Despite multiple further improvements to the premier chess engines over the past few decades, the core algorithm, alpha-beta pruning, remained unchanged until recently—when Deepmind's AlphaZero [9] conquered Stockfish, the leading chess engine at the time. Instead of the traditional alpha-beta pruning algorithm, AlphaZero employed a modified version [8] of *Monte-Carlo Tree Search* (MCTS) [7], which replaces rollouts with a neural network which estimates what a rollout would do. The weights of the neural network were in turn optimised using policy iteration, a form of reinforcement learning, via self-play.

An AlphaZero-inspired open-source engine, *Leela Chess Zero* (Lc0) was developed soon after. Just like AlphaZero, it was trained *tabula rasa*, and briefly surpassed Stockfish. In 2020, however, Stockfish incorporated *efficiently updateable neural network* (NNUE) evaluation[4] (which is a smaller neural network that is optimised for CPUs and is used in conjunction with alpha-beta pruning) and has dominated the Top Chess Engines Competition (TCEC) ever since[5]. As a result, the latest version of Stockfish at this time, Stockfish 16, has been used throughout this research as the 'ground truth' when it comes to evaluating positions and finding the best moves.

Unlike the engines above, the Maia chess engine was developed with the aim of mimicking human players. Maia builds on Lc0—the same executable is used as Lc0, but the neural network weights used by the executable were trained on human games, as opposed to using self-play. The weights for each Maia chess engine were trained on a PGN[6] database comprising 12 million human games

---

[3] In this paper, "depth" refers to the average depth of the MCTS tree, as described in the Leela chess documentation (https://lczero.org/dev/wiki/technical-explanation-of-leela-chess-zero/).

[4] https://stockfishchess.org/blog/2020/introducing-nnue-evaluation/.

[5] https://www.chessdom.com/stockfish-wins-tcec-season-22-sets-records/.

[6] Portable Game Notation: https://www.chessprogramming.org/Portable_Game_Notation.

within an Elo rating[7] band of 100 points [5] obtained from Lichess, an Internet chess server. An engine for each 100-point rating band between 1100 and 1999 was created, resulting in 9 engines in total. The engines performed admirably in terms of move prediction accuracy: accuracy for each Maia rating band ranged between 50–53%, while Stockfish and Lc0 only managed around 35–40% in most cases, with a peak of approximately 46%.

It is important to note that tree search for Maia was disabled in order to obtain optimal results. According to McIlroy-Young et al. [5], using search resulted in engines that were too strong compared to the imitated players, which led to a drop in move-matching accuracy. Additionally, Maia was fed the previous 12 plies[8] before the current board position, and this boosted accuracy by 2–3%.

In a follow-up to the original Maia paper [6], the Maia engines discussed above were tailored for move prediction of individual chess players using transfer learning. These attempts were reasonably successful—the personalised models on average outperformed the closest "generic" Maia model by around 4–5%, however their attempts to create personalised models for GMs (grandmasters) were not nearly as successful as those for players rated below 2000. We hypothesize that this is due to the lack of search, as an engine without tree search is somewhat limited in strength.

In a presentation[9], Reid McIlroy-Young, one of the authors of the papers on Maia, pointed out another drawback of disabling tree search: Maia engines, without search, failed to solve some chess tactics that the majority of players within the respective rating bands would be able to solve. This is, arguably, a bigger problem when it comes to imitating players than it may seem. Incorrect predictions in these critical, tactical positions can have an outsized impact on the outcome of a game, and as a result significantly decrease the quality of the imitations.

Jacob et al. [2] sought to address some of Maia's limitations by implementing a Kullback-Leibler-regularized search, resulting in an improvement of approximately 1% in prediction accuracy for Maia1900. Our proposed modifications attempt to improve further on these promising results.

It should be noted that for at least the last 20 years, attempts have been made at attenuating traditional chess engines so that they are more suitable for sparring play. These efforts include Chessmaster "personalities", Dragon (by Komodo Chess) "personalities", and the plethora of chess.com bots. To the best of the authors' knowledge, these follow the traditional approach of tweaking and attenuating an existing chess engine by adjusting its parameters and by using a custom opening book. While this approach is often perfectly adequate, the Maia framework is more flexible, ambitious, and requires less human intervention. Additionally, this framework could be extended to applications beyond sparring partners, such as the classification of players into stylistic archetypes, cheat

---

[7] https://www.chess.com/terms/elo-rating-chess.

[8] A ply is a single move made by one of the players.

[9] https://www.youtube.com/watch?v=kiFJSgM1d68.

detection, and chess "betting" (since models could be employed to generate odds for moves in live games).

3   Methodology

3.1   Training Our Baseline Engines

In order to thoroughly evaluate the efficacy of our proposed modifications, performance was compared with three baseline engines, from weakest to strongest:

- **Maia1900.** Maia1900 was already trained by McIlroy-Young et al. [5] on 12 million games played between players rated 1900–1999 on Lichess.
- **LC2400.** We trained the LC2400 engine on approximately 8 million games played between players rated 2400 and above on Lichess.
- **OTB2400.** Our strongest engine was trained on roughly 800,000 over-the-board, tournament games by players rated 2400 and above.

The latter two engines were trained from scratch using the same procedure as the Maia engines, as described in the Maia Github repository[10]. In spite of the significantly smaller training set, we did not notice any adverse impact on the engines' performance.

3.2   Experimental Procedure

Our experimental procedure involved comparing a number of experimental models, where each Maia-like engine is modified in some way, to the default performance of the Maia-like engines. Each augmented model generates predictions for the moves in a test set of 1000 games, and two performance metrics (described in the following section) are calculated.

The 1000 games in each test set were collected from publicly-available Lichess data[11] (for the Maia1900 and LC2400 models) and Chessbase Mega Database 2023[12] (for the OTB2400 models). The test sets were constructed as follows:

- The Maia1900 test set comprises 1000 games between players rated 1900–1999 played in December 2019. These games are a subset of the test set used by McIlroy-Young et al. [5].
- The LC2400 test set comprises 1000 games on Lichess played between players rated 2400 and above in July 2022 .
- The OTB2400 test set comprises 1000 tournament games played between players rated 2400 and above during the months of October and November in 2021.

Note that the first 10 plies (5 moves from both sides) were skipped when generating the move predictions, as done in the Maia papers. Additionally, as in the Maia papers, moves where the player had less than 30 s remaining were excluded in the case of Maia1900, since players often tend to become erratic and unpredictable when time gets low.

---

[10] https://github.com/CSSLab/maia-chess.
[11] https://database.lichess.org/.
[12] https://shop.chessbase.com/en/products/mega_database_2023.

3.3   Performance Metrics

In broad terms, when tuning an engine to mimic a set of players, we want the engine to have a similar playing strength (the objective quality of the moves) and a similar playing style (the strategy or logic behind the moves) to the input games. Tuning the strength of the engine is easier than the style, since the former is easier to quantify. The following two performance metrics were considered to represent the notions of style and strength:

- **Prediction accuracy percentage.** This is simply the total number of moves predicted correctly as a percentage of the total number of positions in the test set. It can be seen as both an indicator of playing strength and playing style, and it was the sole metric used in the Maia papers. Naturally, the higher the prediction accuracy, the better.
- **Difference in average centipawn loss (ACPL).** The average centipawn loss is the average absolute difference[13] in evaluation between the best move and the played/suggested move. It is a sensible measure of playing strength, and has been utilised for cheat detection purposes and comparing playing strength [1]. To obtain the difference in ACPL, we subtract the model's ACPL from the actual ACPL. The closer the difference is to zero, the better.

4   Implementation and Results

4.1   Fixed-Depth Search

Since it is evident that one of the primary shortcomings of the baseline Maia-like engines is that they are too weak to accurately mimic master-level players, arguably the most sensible starting point for further investigations is to enable search up to a given depth. MCTS is already implemented within the Lc0 engine executable, meaning that enabling search is straightforward.

This paper uses the notion of "depth" of a search tree, as opposed to the number of rollouts (which is more conventional within the context of MCTS). We use this approach because we feel that "depth" is a more intuitive concept for human chess players. Nevertheless, the models outlined below should not be sensitive to the choice between depth or number of rollouts when limiting search: Fig. 1 illustrates the relationship between depth and the number of rollouts. As shown, the number of rollouts required to increase the average depth of the search tree increases exponentially relative to the depths.

Terminating search, irrespective of whether depth or rollouts are used, is a simple matter through the Universal Chess Interface (UCI) for engines. The top engine move at each depth, from 1 to 7, is recorded, and compared to the actual move in order to obtain that depth's prediction accuracy.

Based on the results in Fig. 2, the optimal search depths for Maia1900, LC2400, and OTB2400 are 4, 6, and 7 respectively. Enabling search up to the

---

[13] It is necessary to average the *absolute* difference since engine evaluations are always from white's perspective.

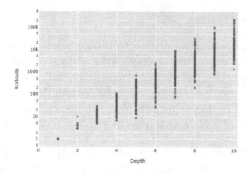

Fig. 1. The relationship between rollouts and search depth for the OTB2400 engine when predictions were made for the positions in 10 games.

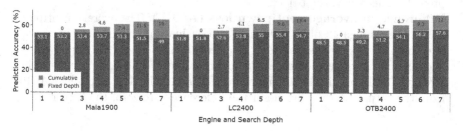

Fig. 2. Fixed depth prediction accuracy.

best depth leads to a considerable improvement in the case of the two stronger baseline engines (LC2400 and especially OTB2400), but it (surprisingly) also slightly improves the prediction accuracy of the Maia1900 engine. The results for the Maia1900 engine stand in contrast to the findings of McIlroy-Young et al. [5], whose experiments indicated that search with 10 rollouts reduced prediction accuracy. Additionally, note that the *cumulative*[14] prediction accuracy at depth 7 is around 12% higher than the best search depth in Fig. 2. This indicates that positions exist where some search depths thrive, and others struggle—this suggests that incorporating a position-dependent depth predictor may be helpful in future.

4.2 Depth Ensemble Model

The diverse predictive behaviour at different search depths suggests that an ensemble model may be effective in improving prediction accuracy. Our implementation of a search depth ensemble was relatively naive: a voting classifier was employed, with each ensemble member's move prediction weighted proportional

---

[14] In other words, the accuracy obtained when calculated using correct move predictions up to a given depth. For instance, the cumulative accuracy at depth 2 includes all the correct predictions at depth 1, as well as the extra correct predictions at depth 2.

to its prediction accuracy on the training set. The highest-weighted move prediction then becomes the ensemble's move prediction, as shown by Algorithm 1.

Algorithm 1. Move Predictions using an Ensemble Model

Input:
    ensembleDepths: A list of depths in the ensemble.
    weights: A list of weights corresponding to each search depth.
    position: The current board position.
Output:
    prediction: The move predicted by the ensemble.

1: moveDictionary ← {}
2: **for** depth in ensembleDepths **do**
3:     move ← ENGINEANALYSIS(position, depth)
4:     moveDictionary[move] += weights[depth]
5: prediction ← GETMAXMOVE(moveDictionary)

Selecting the ensemble members is slightly trickier, as it was found that the inclusion of poor predictors in the ensemble reduced prediction accuracy. Based on some experimentation on a validation set, ensembles that worked well consisted of the three best search depths and any other search depths that were within 2% of the best depth's prediction accuracy in Fig. 2.

4.3   Evaluation Window Model

The evaluation window model takes an entirely different approach, by trying to manage the average centipawn loss metric directly, instead of targeting prediction accuracy. The effect of this approach is shown in Fig. 3, which depicts the kernel density estimate (an estimate of the probability density function) of various models' error sizes. The no-search baseline engine (in orange) is too weak to accurately mimic the actual players (in blue), as evidenced by the relatively high frequency of blunders and relatively low frequency of good moves, while the evaluation window model (in red) is far closer to the playing strength of the actual players.

We use Stockfish 16 (the strongest engine currently available) to stochastically filter out moves so that the strengths of the unfiltered moves are more similar in strength to those typically made by the imitated players. When strengthening the engine, the threshold needs to occasionally filter out poor moves, and vice versa. This is done by randomly sampling a CPL threshold according to an exponential distribution, and then selecting a prediction among moves that do not violate that threshold—see Algorithm 2.

Tuning the exponential distribution for the evaluation window model is a slightly more involved process. First, the average actual CPL and the average CPL of the baseline engine (without search) are calculated and compared—if

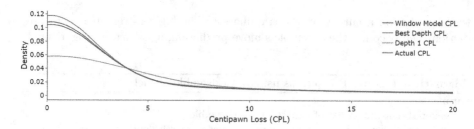

Fig. 3. Kernel density estimate of CPL for OTB 2400 dataset. The upper limit of the x-axis was capped at 20 to make differences between the curves for small values of x clearer.

Algorithm 2. Move Predictions Using an Evaluation Window

Input:
> distribution: An exponential distribution that is tuned during training and from which a CPL threshold is sampled.
>
> strengthen: A boolean indicating if the engine is to be strengthened or weakened.
>
> position: The given board position.

Output:
> move: A final move prediction that is within the CPL threshold.

```
1: sampledCPL ← SAMPLE(distribution)
2: moveList ← BASEMODEL(position)      ▷ Sort predictions from most to least likely.
3: bestEval ← STOCKFISHANALYSIS(position)          ▷ Evaluate the best move.
4: for move in moveList do        ▷ Runs from most to least likely move prediction.
5:     eval ← STOCKFISHANALYSIS(position, move)
6:     cpl ← |bestEval-eval|
7:     if strengthen = True then                ▷ Occasionally reject poor moves.
8:         if cpl < sampledCPL then
9:             return move
10:    else                              ▷ Occasionally reject good moves.
11:        if cpl > sampledCPL then
12:            return move
```

---

the baseline engine's CPL is higher than the actual CPL, then the engine needs to be strengthened (i.e. CPL lowered), and vice versa.

Based on preliminary quantitative distribution fitting and visual inspection of Fig. 3, it seems reasonable to model the CPLs with exponential distributions, which we parameterise by the scale parameter β. In order to strengthen the engine, for instance, the aim is that the evaluation window will filter out bad moves which might otherwise have been popular with the neural network; this would occur when the move's CPL exceeds the threshold. Without proper tuning, the sampled CPL thresholds might be too low (resulting in an engine that is too strong) or too high (resulting in an engine that is too weak).

The core of the algorithm tuning β is given in Algorithm 3, where ϕ denotes the golden ratio. Golden section search [3] is well-suited to this task, since the

---

Algorithm 3. Tuning the Evaluation Window Model

Input:
> games: A PGN database of games.
> threshold: The CPL difference at which training terminates.
> iterations: The number of training iterations.

Output:
> model: A model consisting of the exponential distribution (the scale parameter) and whether the engine is being strengthened/weakened.

1: actualCPL, depth1CPL ← CALCULATECPL(games)   ▷ Uses Stockfish to calculate the CPL of the actual moves and baseline engine's moves.
2: min ← 0   ▷ min represents the minimum of the search space.
3: if actualCPL < depth1CPL then
4:     strengthen ← **True**
5:     max ← 1000   ▷ max represents the maximum of the search space.
6: else
7:     strengthen ← **False**
8:     max ← 5
9: for $n = 1$: iterations **do**
10:     $\alpha \leftarrow \text{max} - \phi(\text{max} - \text{min})$
11:     alphaModel ← GETMODEL(α, strengthen)   ▷ Returns a model that strengthens/weakens the engine using an exponential distribution with scale = α.
12:     alphaCPL ← EVALUATEMODEL(alphaModel, games, strengthen)   ▷ Uses Stockfish to calculate the CPL of the augmented model.
13:     fAlpha ← |actualCPL - alphaCPL|
14:     if fAlpha < threshold then
15:         return alphaModel
16:     $\beta \leftarrow \text{min} + \phi(\text{max} - \text{min})$
17:     betaModel ← GETMODEL(β, strengthen)
18:     betaCPL ← EVALUATEMODEL(betaModel, games, strengthen)
19:     fBeta ← |actualCPL - betaCPL|
20:     if fBeta < threshold then
21:         return betaModel
22:     if fAlpha < fBeta then
23:         max ← β
24:     else
25:         min ← α

---

function to be optimised is expensive to evaluate and is not differentiable. The algorithm, in essence, iteratively reduces the search space using the golden ratio until an optimum is found. Note that the algorithm is highly dependent on the function evaluations being reliable, which is why it is so vital to use many games for training (in this case, 1000).

5   Discussion

This section compares and discusses the results of the various models, and draws conclusions based on these results. The results are summarized in Table 1. For each engine-metric combination, one point was awarded to a model (shown in bold) that was best in that particular setting[15]. The points are tallied up in the "score" rows in order to make the comparison of the models easier.

Table 1. Summary of Results. "D1" refers to depth 1 (no search), "BD" stands for best depth, "EM" is short for ensemble model, and "EW" represents the evaluation window model.

Metric	Engine	D1	BD	EM	EW
Accuracy	Maia1900	53.4%	54.1%	**54.6%**	53.0%
	LC2400	52.7%	55.4%	**55.9%**	54.2%
	OTB2400	49.3%	**58.4%**	**58.4%**	56.4%
	Subtotal Score	0	0.5	**2.5**	0
ACPL Difference	Maia1900	−2.1	−21.4	−16.9	**−0.7**
	LC2400	19.1	−9.5	−9.8	**−1.6**
	OTB2400	32.3	3.4	4.6	**−0.1**
	Subtotal Score	0	0	0	**3**
TOTAL SCORE		0	0.5	2.5	**3**

According to these scores, the evaluation window approach narrowly beat the ensemble model. The best model for deployment, however, depends heavily on which metric is deemed most important for the application. If prediction accuracy is prioritised, then the ensemble model is to be preferred, as it achieved the highest accuracy for all three engines.

If the priority is to ensure the strength of the model is as close to the actual players' as possible, however, then the evaluation window model appears to be a better choice. Since this model optimises for ACPL, it is unsurprising that it performed best in all cases using the ACPL difference metric. Additionally, it beats the accuracy of the baseline engine for the OTB2400 and LC2400 datasets, but falls short of the best-depth or ensemble models.

Similarity in playing strength does not guarantee the highest move prediction accuracy, as is evident when contrasting the best-depth and evaluation window models. In all three datasets, the ACPL difference dramatically improved, yet the prediction accuracy dropped. This suggests (or rather it confirms) that there are other factors at play, such as style. Based on this, it appears as though a trade-off exists when seeking to imitate chess players—aligning playing strength may come at the cost of reducing stylistic similarity, and vice versa.

---

[15] In the case of a tie, the point was shared.

Fig. 4. A puzzle taken from a game between Ftáčnik and Vallejo Pons in 2007.

The downside of weighing "style" too heavily is perhaps most evident in tactical positions, where one move is clearly superior. This trade-off is illustrated in Fig. 4, which shows a position from a game between two grandmasters, Ftáčnik and Vallejo Pons. In the featured position, the white player has a relatively simple continuation (indicated using the green arrow), leading to checkmate in three moves (43.Qg8+ Kxg8 44.Bd5+ Kf8 45.Rg8#). Since it is exceptionally likely that most master-level players would find this sequence, it is logical to compare the performance of the various OTB2400 models. The evaluation window model was the only one which found the correct continuation, whereas the baseline engine predicted 43.Be4 (orange arrow), and both the best depth and depth ensemble models opted for 43.Bf1 (red arrow) instead. Both erroneous predictions are refuted by 43...Rg6.

Such oversights are unusual for master-level players. While inaccurate predictions in these complex, tactical positions may not significantly affect overall prediction accuracy, their impact on the quality of imitation is disproportionate due to the magnitude of the error. This reinforces the need to account for the objective strength of predicted moves when assessing the quality of mimetic models of chess players, particularly stronger ones.

6   Conclusions and Future Work

This research set out to create models that mimic human chess players more accurately than existing approaches, both in terms of playing style and strength. In particular, the goal was to develop human-like models that were able to imitate master level players.

The Maia chess training infrastructure was used to train two models, named LC2400 and OTB2400, on the games of players rated 2400 and above on Lichess and over-the-board respectively. A number of modifications to these models (as well as Maia1900, an existing Maia engine) were developed, tested and then evaluated using two metrics: prediction accuracy and average centipawn loss. It was found that the ensemble model, which combined predictions from an ensemble of models based on different search depths, achieved the highest prediction accuracy, while the evaluation window model, which filters out certain candidate moves based on external objective evaluation of the moves, performed best with respect to the average centipawn loss difference metric.

Future work could entail applying the approaches to different groups of players, developing a "depth predictor" based on the position, improving the optimisation algorithm for training the evaluation window, exploring the potential strength-style trade-off, and creating personalised models for individual players.

References

1. Ferreira, D.: Determining the strength of chess players based on actual play. ICGA J. **35**, 3–19 (2012)
2. Jacob, A.P., et al.: Modeling strong and human-like gameplay with KL-regularized search. In: Chaudhuri, K., Jegelka, S., Song, L., Szepesvari, C., Niu, G., Sabato, S. (eds.) Proceedings of the 39th International Conference on Machine Learning. Proceedings of Machine Learning Research, vol. 162. PMLR, 17–23 July 2022. https://proceedings.mlr.press/v162/jacob22a.html
3. Kiefer, J.: Sequential minimax search for a maximum. Proc. Am. Math. Soc. **4**(3) (1953). http://www.jstor.org/stable/2032161
4. McCarthy, J.: Chess as the drosophila of AI. In: Marsland, T.A., Schaeffer, J. (eds.) Computers, Chess, and Cognition. Springer, New York (1990). https://doi.org/10.1007/978-1-4613-9080-0_14
5. McIlroy-Young, R., Sen, S., Kleinberg, J., Anderson, A.: Aligning superhuman AI with human behavior: chess as a model system. In: Proceedings of the 26th ACM SIGKDD International Conference on Knowledge Discovery and Data Mining. KDD '20, Association for Computing Machinery, New York, NY, USA (2020). https://doi.org/10.1145/3394486.3403219
6. McIlroy-Young, R., Wang, R., Sen, S., Kleinberg, J., Anderson, A.: Learning models of individual behavior in chess. In: Proceedings of the 28th ACM SIGKDD Conference on Knowledge Discovery and Data Mining. KDD '22. Association for Computing Machinery, New York, NY, USA (2022). https://doi.org/10.1145/3534678.3539367
7. Munos, R.: From bandits to Monte-Carlo tree search: the optimistic principle applied to optimization and planning. Found. Trends® Mach. Learn. **7**(1) (2014)
8. Rosin, C.: Multi-armed bandits with episode context. Ann. Math. Artif. Intell. **61** (2010). https://doi.org/10.1007/s10472-011-9258-6
9. Silver, D., et al.: A general reinforcement learning algorithm that masters chess, shogi, and go through self-play. Science **362**(6419) (2018). https://doi.org/10.1126/science.aar6404

Merging Neural Networks with Traditional Evaluations in Crazyhouse

Anei Makovec[1], Johanna Pirker[2], and Matej Guid[1(✉)]

[1] University of Ljubljana, Ljubljana, Slovenia
`matej.guid@fri.uni-lj.si`
[2] Graz University of Technology, Graz, Austria

Abstract. In the intricate landscape of game-playing algorithms, Crazyhouse stands as a complex variant of chess where captured pieces are reintroduced, presenting unique evaluation challenges. This paper explores a hybrid approach that combines traditional evaluation functions with neural network-based evaluations, seeking an optimal balance in performance. Through rigorous experimentation, including self-play, matchups against a variant of the renowned program, Go-deep experiments, and score deviations, we present compelling evidence for the effectiveness of a weighted sum of both evaluations. Remarkably, in our experiments, the combination of 75% neural network and 25% traditional evaluation consistently emerged as the most effective choice. Furthermore, we introduce the use of Best-Change rates, which have previously been associated with evaluation quality, in the context of Monte Carlo tree search-based algorithms.

Keywords: Crazyhouse · chess variants · heuristic evaluation functions · neural networks · Best-Change rates · Monte Carlo tree search

1 Introduction

Crazyhouse, a chess variant, introduces a unique gameplay element where players can place captured pieces back on the board as their own. This rule significantly increases the game's branching factor and state space complexity compared to standard chess. The strategies in Crazyhouse also differ: defensive moves, especially castling, become crucial due to increased threats to the king. Additionally, the typical values assigned to the chess pieces become less predictable due to the potential of held pieces [2].

Historically, the approach to mastering Crazyhouse relied heavily on conventional chess-playing techniques. These included heuristic search algorithms enhanced by Alpha-Beta pruning and evaluation functions deeply rooted in game-specific knowledge. However, the introduction of AlphaZero shifted this

© The Author(s), under exclusive license to Springer Nature Switzerland AG 2024
M. Hartisch et al. (Eds.): ACG 2023, LNCS 14528, pp. 15–25, 2024.
https://doi.org/10.1007/978-3-031-54968-7_2

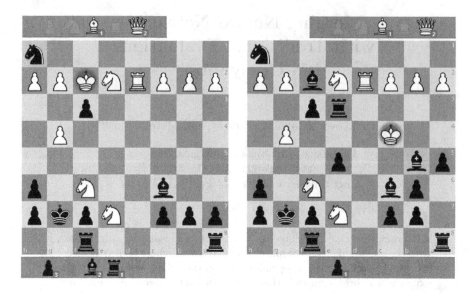

Fig. 1. An example of an attack in Crazyhouse. On the left, Black has just reintroduced a previously captured knight onto the board with the move 1...N@h1+, placing the white king in check. Note that White has a bishop and two queens in hand, while Black possesses three pawns, two bishops, and a rook as held pieces that play the crucial role in the attack. Following the sequence 2.Ke3 B@f2+ 3.Kd3 R@e3+ 4.Kd4 @e5+ 5.Kc5 @b6+ 6.Kb4 a5+ 7.Kc4 B@b5#, the position on the right emerges. The white king is checkmated.

paradigm. This algorithm combined deep reinforcement learning with Monte Carlo tree search, mastering games like chess, shogi, and Go using only self-play, thus reducing the apparent need for domain knowledge [10]. This development led to a new generation of game-playing programs that heavily emphasized deep learning, sometimes setting aside or even excluding the traditional domain knowledge previously considered as essential.

Despite the rise of neural networks in Crazyhouse, the value of traditional programs, built on years of domain knowledge, remained clear [1]. In this article, we introduce CRAZYRABBIT, our program that was built with a sole purpose: to explore a hybrid approach, namely the use of domain knowledge in conjunction with deep learning, when it comes to playing the game of Crazyhouse. CRAZYRABBIT's unique feature is the hybrid evaluation function, designed to merge neural-network-based evaluations with traditional game-specific knowledge. To test the effectiveness of this approach, we conducted four experiments: self-play experiment, matchups against a handicapped version of the well-known FAIRY-STOCKFISH program, the Go-deep experiment, and score deviations experiment.

A hybrid approach between traditional and neural network game-playing techniques is not common practice. He et al. [6] combined a traditional evaluation function with a neural network evaluation to play Chinese chess, using a weighted sum of their evaluations. Another example of such an approach comes from Yan and Feng [13], where for the game of Gomoku, a hard-coded evaluation function was combined with a neural network, again using a weighted sum.

Figure 1 shows a typical and well conducted attack by black pieces. Both players have a set of pieces beside the chessboard, which can be reintroduced during their respective turns. In this instance, White's two queens in hand remain unused, as Black's successive checks only allow the king to move until the eventual checkmate. The moves are given in standard algebraic notation, with the "@" sign indicating moves using pieces that were previously off the board.

The article is structured as follows. We first explain how the evaluation function was constructed and introduce our hybrid approach. This is followed by a detailed examination of each of the four experiments aimed at evaluating CRAZYRABBIT's performance. We then conclude with a synthesis of the results and their implications for future game-playing algorithm evaluation.

2   Implementation of CrazyRabbit

In developing CRAZYRABBIT, we have largely relied on the design decisions of AlphaZero algorithm [10] and CRAZYARA [1] as they describe the current state of the art for playing Crazyhouse using neural networks. For move generation, we used SURGE[1], a bitboard-based legal chess move generator, and modified it to support Crazyhouse's dropping moves.

Our primary research goal was to investigate the integration of traditional evaluation functions with neural network evaluations within game-playing contexts. Achieving amateur-level strength was considered adequate for this purpose. The full source code of our program is available on *GitHub*[2].

2.1   Monte Carlo Tree Search

To implement Monte Carlo tree search (MCTS), we used an approach similar to that used in CRAZYARA, which is based on the version of the PUCT algorithm from AlphaZero [10]. However, we did not use any of the additional modifications to the original algorithm proposed in CRAZYARA, and our version was single-threaded.

2.2   Deep Neural Network

For the neural network model architecture of our program, we chose the architecture proposed and used in CRAZYARA. It still uses a dual architecture design with a tower of residual blocks followed by a value and policy head proposed in AlphaZero [10], but it differs from the original design by using 13 inverted

---

[1] https://github.com/nkarve/surge.
[2] https://github.com/AneiMakovec/CrazyRabbit.

residual blocks [8] with group depthwise convolutions instead of the default ones. It also uses the concept of the Pyramid-Architecture [5], in which the number of channels for the 3×3 convolutional layer of each consecutive residual block is increased. According to Czech et al. [1] this is done because after the first convolution layer only around half activation maps are used, which is due to the more compact input representation.

2.3   Training Data and Supervised Learning

We created a dataset consisting of ranked games played by human players on the online platform *Lichess*[3] from July 2021 to February 2022. To maximize the training potential of our dataset, we selected only games in which both players had an Elo ≥ 2000 and which were not aborted. This brought the size of our dataset to 316,950 games.

The average player Elo for a single game in our dataset was 2203.49, and short time control game modes predominated, with 42.63% of all games being one-minute games. The popularity of short games could potentially affect training, as 27.13% of all games ended with a time forfeit. The quality of moves usually decreases over the course of a game, especially when the player is under pressure because of the time running out. On the other hand, the games were played by many different players, with the most frequent player participating in 2.73% of all games. This means that our dataset was broader and could contain many different game strategies that the model could learn.

We used the same approach as Czech et al. [1] and trained the neural network on the dataset with supervised learning, using the same parameters as they did.

2.4   Evaluation Function and Hybrid Approach

To implement our evaluation function, we relied on the implementations of the chess playing programs REBEL [9] and ROOKIE 2.0 [12], as their implementation decisions and design considerations are described in detail. It consisted of five different features: material advantage, pawn structure, king safety, piece placement and board control. The only thing we changed were the piece values of the material advantage feature, because compared to standard chess in Crazyhouse we also have to consider the value of the pieces in the hand. For this, we used the values recommended by Droste and Fürnkranz [2], which were developed specifically for Crazyhouse using reinforcement learning.

Traditional evaluation functions usually represent their evaluation in centipawns, a unit that measures the pawn advantage that one player has over the other, whose value is unbounded and can be both positive and negative. On the other hand, neural networks used by programs based on AlphaZero usually use the winning percentage to describe how likely a player is to win, and its value is limited to the interval $[-1, 1]$.

---

[3] https://lichess.org/.

To use a traditional evaluation function in conjunction with a neural network, we must consider the different units in which each approach represents the evaluated value and convert them into the same metric. We did this using the equation used in the LCZERO[4] program:

$$cp = 111.714640912 \cdot tan(1.5620688421 \cdot v), \tag{1}$$

for which we find the inverse and get:

$$v = 0.64018 \cdot arctan(0.00895 \cdot cp), \tag{2}$$

where cp and v represent the evaluations returned by the evaluation function and neural network, respectively.

The evaluation function we used calculates its evaluation score in centipawns, which are then converted to winning percentages using Eq. 2 before being returned. Like our neural network model, it always returns the score from the perspective of the current player. Then, the score of the neural network ($value_{nnet}$) is combined with that of the evaluation function ($value_{eval}$) by a weighted sum:

$$value_{final} = \alpha \cdot value_{nnet} + (1 - \alpha) \cdot value_{eval}, \tag{3}$$

where α is the weighting coefficient, with a value in the range $[0, 1]$. However, since α can be set arbitrarily, we also focused on finding the best value for it in our experiments.

3   Self-play Experiment

We conducted a series of self-play matches, with each match consisting of 100 games. For these games, both participating programs were set to use 1000 simulations per move, while varying the α weights for their evaluations.

Figure 2 shows the results from the perspective of the program with $\alpha = 0.75$. This particular version proved to be superior, outperforming all other variations. We evaluated six different program versions based on different α values: $[0.10, 0.25, 0.50, 0.90, 1.00]$. Notably, the program with $\alpha = 1.00$ used only the neural network evaluations and completely bypassed the traditional evaluation function.

Each bar in the graph represents the relative Elo rating, each supplemented by the 95% confidence interval. The version with $\alpha = 0.75$ stands out, consistently outperforming the other versions with considerable statistical significance. The version with $\alpha = 0.50$ is relatively close behind and shows similar results to the version with $\alpha = 0.75$. This similarity is also evident in our Go-deep and score deviations experiments (see Sects. 5 and 6).

---

[4] https://github.com/LeelaChessZero/lc0.

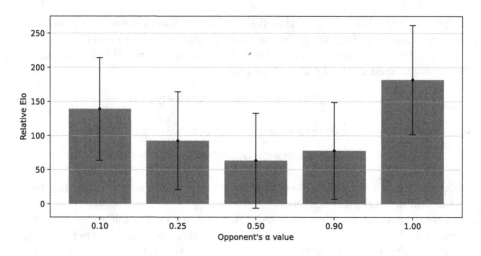

Fig. 2. Results of self-play experiments, where our program with $\alpha = 0.75$ played against other versions. All matches consisted of 100 games and both players used 1000 simulations per move.

4   Experiment with Fairy-Stockfish

To get a feel for how our program would perform in practice, we had it play against FAIRY-STOCKFISH[5], a variant of the popular chess program STOCKFISH[6]. Default settings were used for FAIRY-STOCKFISH, with the exception of UCI_Elo and UCI_LimitStrength, which were adjusted to set a specific Elo strength.

Each match consisted of 200 games, and our program was set to use a fixed number of simulations per move, while FAIRY-STOCKFISH was limited to 2 s per move to save time.

Figure 3 shows the results of our program using 1000 simulations per move against FAIRY-STOCKFISH at 1600 Elo. Our program was not able to beat FAIRY-STOCKFISH using the neural network alone ($\alpha = 1.00$), but it was able to beat it using our hybrid approach ($\alpha = 0.75$).

---

[5] https://github.com/ianfab/Fairy-Stockfish.
[6] https://github.com/official-stockfish/Stockfish.

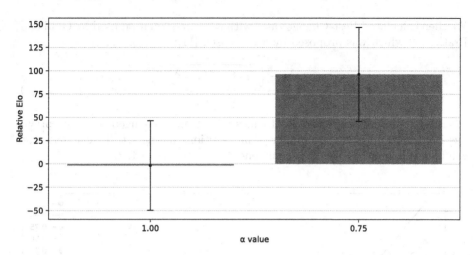

Fig. 3. Results of CRAZYRABBIT with 1000 simulations per move playing against FAIRYSTOCKFISH set at 1600 Elo and with 2 s per move.

5 Go-Deep Experiment

In the Go-deep experiments we are interested in how the best moves provided by the program for a given set of positions change with respect to search depth. The approach is based on Newborn's observation [7] that the results of self-play experiments are closely correlated with the rate at which the best move changes from one iteration to the next. We are interested in a comparison of the Best-Change rates of the programs.

Guid and Bratko [3] observed that Best-Change rates depend on (1) the values of the positions in the dataset and (2) the quality of the evaluation function of the program used. They noted that Best-Change rates might reflect the quality of a program's evaluation function. Specifically, stronger evaluation functions tend to change their preferred move less frequently as search depth increases. This led to the suggestion that Best-Change rates might directly measure the quality of a evaluation function (under certain circumstances that require further investigation). However, this idea has not yet been explored in Monte Carlo tree search-based programs.

To build the set of positions for our experiment we used the same set of games as we used to train the neural network in Sect. 2.3. We selected 10,000 random unique positions that were reached after 12 half-moves (plies) or more and which our neural network evaluated as approximately balanced (within the interval of $[-0.01, 0.01]$).

Since AlphaZero-based programs do not measure their search depth in the traditional sense (fully explored depths) but rather in the number of simulations performed, we ran our program on the position set with different numbers of simulations (from 100 to 1000) and observed in how many positions the best move changed from the one returned with the previous number of simulations.

We then calculated Best-Change rates with the same equation used by Guid and Bratko [3], which was proposed by Steenhuisen [11]:

$$\frac{m + \frac{\lambda^2}{2} \pm \sqrt{m(1 - \frac{m}{n}) + \frac{\lambda^2}{4}}}{n + \lambda^2}, \tag{4}$$

where m represents the number of times a best move changed from the previous iteration and n is the number of all positions in the set. To calculate the confidence bounds we use 95% confidence, by setting λ to 1.96.

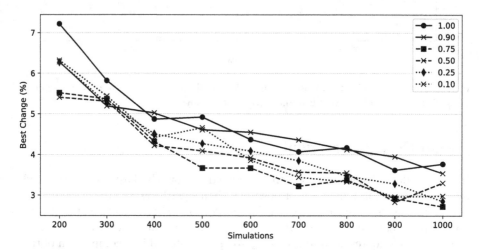

Fig. 4. Results of the Go-deep experiments on 10,000 selected positions with different numbers of simulations per move for different α values.

Figure 4 depicts the Best-Change rates of our program across various α values. The hybrid approach ($\alpha < 1.00$) consistently demonstrates lower Best-Change rates than the neural network-only approach ($\alpha = 1.00$). Specifically, at 1000 simulations, the rate for $\alpha = 0.75$ is 2.72 ± 0.16, compared to 3.77 ± 0.19 for $\alpha = 1.00$. Notably, $\alpha = 0.75$ shows statistically significant lower rates than $\alpha = 1.00$ for all simulation counts in the experiment.

6   Average Move Score Deviations

Guid and Bratko [4] proposed a heuristic-search based approach to estimating skill levels in game playing. Chess players were benchmarked against each other based on their scores, calculated as average differences in the evaluation values between the moves actually played and the best moves found by the program. They used different chess programs to perform analyses of large data sets of recorded human decisions, and obtained very similar rankings of skill-based performances of selected chess players using any of these programs at various levels of search.

We adopted this method to evaluate the variants of our program. Essentially, we used an independent evaluation function to compare the evaluation of the moves chosen by our program with the evaluation of the optimal moves.

In the experiment, we used FAIRY-STOCKFISH, and its evaluation at a search depth of 12 as an oracle. We then calculated the average score deviation between the best move chosen by FAIRY-STOCKFISH and the move chosen by our program for the same set of 10,000 positions used in our Go-deep experiments. Positions were disregarded if FAIRY-STOCKFISH evaluated either move in terms of moves to mate. We also limited the scoring difference to 300 centi-pawns to account for the unbounded nature of the centi-pawn scoring and to prevent larger deviations from distorting the average results.

Figure 5 shows the average score differences obtained by our program for different α values. Once again, it can be seen that our hybrid approach performs better than using the neural network alone, especially for higher simulation values where better moves are predominantly selected. The α value of 0.75 consistently proves to be optimal, as it achieves the lowest average score differences compared to all other values.

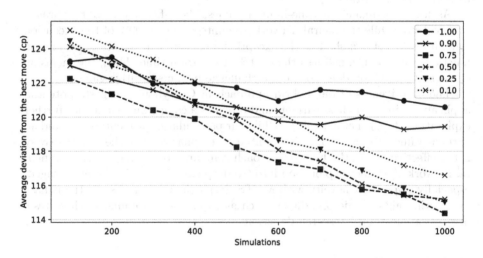

Fig. 5. Average score differences between the best and chosen move with different numbers of simulations in 10,000 selected positions for different α values.

7   Conclusion

In our research, we aimed to combine traditional evaluation functions with neural network evaluations for Crazyhouse game-playing algorithms. We achieved this by using a weighted sum of both evaluations. Using this combined approach, our program CRAZYRABBIT, which employed the Monte Carlo tree search, was tested in four distinct experiments: self-play, matchups against a handicapped version of FAIRY-STOCKFISH, Go-deep experiment and score deviations.

Across all four experiments, the combination that allocated 75% to the neural network and 25% to the traditional evaluation function consistently showed the best performance, highlighting the potential of our hybrid approach.

Crafting an effective evaluation function for Crazyhouse is challenging, especially given its unique mechanic of reintroducing captured pieces. This has led to questions about the most suitable methods to evaluate these functions.

Guid and Bratko's Go-deep experiments [3] highlighted that changes in the best move with deeper searches (termed "Best-Change rates") correlate with the quality of a program's evaluation function. This prompted questions about the viability of Best-Change rates as a quality metric. Our study marks the first exploration of this metric in the context of Monte-Carlo Tree Search-based programs. Our results confirm its utility but it is essential to be aware of its limitations. While successful programs might display traits that align with Best-Change rates as a valid comparative measure, caution is needed. For example, a hypothetical program picking moves alphabetically would have a consistent Best-Change curve, regardless of its inefficacy. Hence, the metric should be used judiciously, and further experimentation may be beneficial to fully understand its scope and limitations.

As for performance in time-limited games, the efficiency of our approach was evident. While the neural network accounted for over 90% of the execution time, it could not overshadow the exceptional efficiency of the static evaluation function, which consumed less than 0.1% of the total time. Thus, the inclusion of the static function does not pose a challenge in time-sensitive games.

The effectiveness of our hybrid approach in Crazyhouse suggests potential applicability in other chess variants and provides a compelling avenue for further exploration of its generalizability. While our particular implementation found an optimal value of $\alpha = 0.75$, it is essential to emphasize that the key takeaway is the effectiveness of the hybrid approach compared to relying solely on a neural network. A deeper exploration into Go-deep experiments is also encouraged, especially given their novelty with MCTS. In particular, contrasting the behavior of other algorithmic paradigms, such as Alpha-Beta pruning, in these tests presents an intriguing research direction.

References

1. Czech, J., Willig, M., Beyer, A., Kersting, K., Fürnkranz, J.: Learning to play the chess variant Crazyhouse above world champion level with deep neural networks and human data. Front. Artif. Intell. **3**, Article 24 (2020). Frontiers Media SA
2. Droste, S., Fürnkranz, J.: Learning the piece values for three chess variants. In: ICGA J. **31**, 209–233 (2008). IOS Press
3. Guid, M., Bratko, I.: Factors affecting diminishing returns for searching deeper. ICGA J. **30**(2), 75–84 (2007)

4. Guid, M., Bratko, I.: Using heuristic-search based engines for estimating human skill at chess. ICGA J. **34**(2), 71–81 (2011)
5. Han, D., Kim, J., Kim, J.: Deep pyramidal residual networks. In: Proceedings of the IEEE Conference on Computer Vision and Pattern Recognition, pp. 5927–5935 (2017)
6. He, Y., Wang, X., Fu, T.: A combined position evaluation function in Chinese chess computer game. Trans. Comput. Sci. **XVII**, 31–50 (2013). Springer
7. Newborn, M.: A hypothesis concerning the strength of chess programs. ICGA J. **8**(4), 209–215 (1985)
8. Sandler, M., Howard, A., Zhu, M., Zhmoginov, A., Chen, L.C.: MobileNetV2: inverted residuals and linear bottlenecks. In: Proceedings of the IEEE Conference on Computer Vision and Pattern Recognition, pp. 4510–4520 (2018)
9. Schröder, E.: How REBEL plays chess (2002)
10. Silver, D., et al.: A general reinforcement learning algorithm that masters chess, shogi, and Go through self-play. Science **362**, 1140–1144 (2018). American Association for the Advancement of Science
11. Steenhuisen, J.R.: New results in deep-search behaviour. ICGA J. **28**(4), 203–213 (2005)
12. Van Kervinck, M.N.J.: The design and implementation of the Rookie 2.0 chess playing program. Master's thesis, Technische Universiteit Eindhoven (2002)
13. Yan, P., Feng, Y.: Using convolution and deep learning in Gomoku game artificial intelligence. Parallel Process. Lett. **28**(03), 1850011 (2018)

Stockfish or Leela Chess Zero?
A Comparison Against Endgame Tablebases

Quazi Asif Sadmine, Asmaul Husna(✉), and Martin Müller⬤

University of Alberta, Edmonton, Canada
{sadmine,asmaul,mmueller}@ualberta.ca

Abstract. The game of chess has long been used as a benchmark for testing human creativity and intelligence. With the advent of powerful chess engines, such as Stockfish and Leela Chess Zero (Lc0), endgame studies have also become a tool for evaluating the capabilities of machine chess engines. In this work, we conduct a detailed study of Stockfish and Lc0, two leading chess engines with distinct methods of play, using chess endgames with varying numbers of remaining pieces. We evaluate the programs' move decision errors when using only the raw policy network as well as when using a small amount of search. We provide insights into the strengths and weaknesses of Stockfish and Lc0 in handling complex endgame positions by exploring common mistakes and identifying interesting behaviours of the engines based on the position of the opponent's last pawn remaining on the board.

Keywords: Computer Chess · Leela Chess Zero · Stockfish · Chess Endgame Tablebases

1 Introduction

Playing chess requires strategic thinking, planning, and decision-making skills. Endgame studies, which involve analyzing and solving complex chess positions with a limited number of pieces remaining on the board, have traditionally been used to test human creativity and intelligence. Endgame studies have also become a tool for evaluating the capabilities of chess engines [9,12].

Before the widespread adoption of deep neural networks and the emergence of AlphaZero [18,19], Stockfish was the leading chess engine. AlphaZero [19] demonstrated superhuman performance in complex board games - chess, shogi, and Go. This neural network-based program has exceptional move selection and state evaluation abilities. Inspired by the success of AlphaZero, Stockfish incorporates a neural network known as NNUE (efficiently updatable neural network) [16] into its traditional chess engine from version 12. However, NNUE is a relatively simple and shallow feedforward neural network, whereas AlphaZero uses a more complex and deep convolutional neural network (CNN).

Q.A. Sadmine and A. Husna—Equal contribution.

© The Author(s), under exclusive license to Springer Nature Switzerland AG 2024
M. Hartisch et al. (Eds.): ACG 2023, LNCS 14528, pp. 26–35, 2024.
https://doi.org/10.1007/978-3-031-54968-7_3

Despite their remarkable performance, these programs are not perfect and still make mistakes. To better understand how these modern programs learn to play as well as explore the limits of their playing abilities, we turn to a sample problem that has known exact solutions - chess endgames. While the full game of chess has not been solved, exact solutions for endgames with up to seven pieces have been computed and compiled into endgame tablebases. In this study, we utilize the open-source programs Leela Chess Zero (Lc0) [1], which follows the AlphaZero-style approach, and Stockfish [4] to analyze chess endgames and investigate the gap between strong and perfect play. Additionally, we compare their respective performance in these endgames. Through this analysis, we aim to shed light on their playing abilities and provide insights into their strengths and limitations. We design a comprehensive methodology involving extensive experiments to address the following research questions:

- How well do these two leading chess engines perform compared to perfect play?
- Which is easier to predict for the engines? Wins or draws?
- Which engine's policy networks perform well in evaluation?
- How much do the programs improve after using a small search budget?
- In an interesting board configuration, when only one pawn of the opponent remains, how much do their performances differ from each other?

2 Related Work

The original AlphaZero paper [19] compared the performance of AlphaZero with Stockfish for chess in terms of gameplay. The authors compared win-draw-loss percentage against the baselines in a tournament under the same time settings, and AlphaZero outperformed Stockfish. However, it's worth noting that NNUE had not been introduced into Stockfish at that time.

The work by Haque et al. [9] compares Leela Chess Zero with perfect play from endgame tablebases. The authors also analyze different case studies of Lc0's policy and search, which give more insights into the performance. In our study, we conduct a detailed analysis with more complex endgames and do further experiments where the engines tend to make more mistakes.

Two noteworthy papers in the field of comparing game engines or algorithm performance to perfect play are worth mentioning. Lassabe et al. [12] use genetic programming to solve chess endgames by combining elementary chess patterns defined by domain experts. In Romein and Bal [17], the game of awari was first solved and then used as a basis to measure the performance of two world champion-level engines from the 2000 Computer Olympiad.

3 Background

3.1 Endgame Tablebases

Chess endgames are sub-problems that occur when only a reduced set of game pieces remain on the board, and the full rules of chess still apply. The solutions are publicly available in databases known as endgame tablebases [14]. Each solution in the tablebase includes the outcome of the game assuming perfect play from both players, along with the optimal moves for reaching that outcome and specific metrics such as the number of plies required to achieve the result. Endgame tablebases hosted online differ in storage size and metrics [11,13].

Tablebase generators are also available and allow for the creation of custom endgame tablebases. Among the available options, Syzygy [13] and Gaviota [5] tablebases are popular and widely used, and they are also free for public access.

3.2 Stockfish

Stockfish is a highly robust open-source chess engine. It takes a position on the chessboard as input and generates a move as output using an alpha-beta pruning search algorithm [7]. To cope with the vast search space of chess, Stockfish employs techniques such as forward pruning and reduction to reduce the search space [10]. The evaluation function of Stockfish determines whether a leaf node is favourable for White or Black by evaluating factors such as the current positions of the pieces, piece activity, game phase, etc.

From version 12, Stockfish uses an efficiently updatable neural network (NNUE) [16] as its evaluation function. This neural network is capable of predicting the output of the evaluation function at a moderate network depth. The architecture of NNUE is shallow, consisting of four layers, and is specifically optimized for speed on the CPU. NNUE has greatly enhanced the performance of Stockfish, making it even more powerful in analyzing chess positions and generating strong moves.

3.3 Leela Chess Zero

Leela Chess Zero (Lc0) is a chess adaptation of the popular Go program Leela Zero. Both open-source programs aim to replicate the success of AlphaZero in their respective games [2]. Similar to AlphaZero [19], Lc0 takes a sequence of consecutive raw board positions as input and utilizes a two-headed network for policy and value estimation. It uses Monte Carlo Tree Search (MCTS) as a search algorithm [6] to find the best move. Over time, the Lc0 developers introduced enhancements that were not present in the original AlphaZero, including additional auxiliary outputs such as the "moves left" head, which predicts the number of plies remaining in the current game [8]. Another auxiliary output called the "WDL head" separately predicts the probabilities of winning, drawing, or losing the game [3].

The raw network policies of Stockfish and Lc0 are very different from each other. It is challenging for humans to comprehend what is happening inside these networks [15]. However, comparing the move decisions of these engines provided solely by the network can help to understand how well they have learned to evaluate endgame positions.

4   Dataset and Evaluation Method

4.1   Dataset Generation and Preprocessing

We follow the dataset generation and preprocessing technique of Kryukov used [11] in [9]. We first place the kings on the board in all possible cases. Then we place other pieces one at a time to generate all positions with three, four, and five pieces. We apply colour swapping, horizontal mirroring, vertical mirroring, and diagonal mirroring for a more efficient generation. Then we remove illegal positions such as pawns on promotion ranks, positions where the player to move can capture the king, etc. We disable castling and en passant captures for all positions for simplicity. We also set the halfmove clock to 0 and the fullmove counter to 1. We obtain all unique legal positions of an endgame and append the *syzygy* endgame tablebase's perfect information for each position. Due to the abundance of five-piece chess endgame positions and limited resources, we sample 1% of all five-piece positions. We consider a total of four three-piece, seven four-piece, and six five-piece tablebases for our experiment. The three-piece tablebases consist of one white queen, rook, knight, or bishop, and both kings. In four-piece tablebases, we include two pieces for each player to maintain a balanced power dynamic between the two sides. In the five-piece tablebases, we position only one pawn for black while white receives any two pieces among the queen, rook, knight, and bishop.

We generate all the positions as Forsyth-Edwards Notation (FEN)[1] strings in our datasets. White pieces are represented by capital letters, and black pieces are represented as lowercase letters in each FEN string. We name our datasets accordingly. For instance, the dataset *KQkp* has a white king, a white queen, a black king, and a black pawn.

We store all positions in a MySQL database and separate the data for each tablebase into two parts: (i) positions that result in a win and (ii) positions that result in a draw. We do this separation to analyze deeply if the engines have any performance variation for winning or drawing positions. We skip losing positions because they are of no use in identifying mistakes made by the engines.

4.2   Engine Settings

For our experiment, we consider the latest versions available at the time of this work, which are Stockfish 15.1[2] and Lc0 0.29.0[3]. To fairly compare the engines, we ensure that they have similar strengths. The highest Elo rating available for Stockfish at the time of this work is 2850. Therefore, we use this Elo rating for both engines. Initially, we focus on comparing only the policy. As the backend, we use CPU for Stockfish since it only runs on CPU. For Lc0, we use *cudnn* on a Linux machine equipped with an Nvidia Titan RTX GPU.

---

[1] https://www.chessprogramming.org/Forsyth-Edwards_Notation.
[2] https://stockfishchess.org/blog/2022/stockfish-15-1.
[3] https://github.com/LeelaChessZero/lc0/releases/tag/v0.29.0.

5   Experimental Results

5.1   Wrong Play Analysis

Table 1 displays the percentage of mistakes made by the raw policies of Stockfish and Lc0 when evaluating move decisions for three, four, and one five-piece endgame tablebases. Mistakes are defined as moves that change the game-theoretic outcome. When there is a winning move available in the endgame tablebase, a mistake is counted if the engine suggests its best move that results in a draw or loss. When there is a drawing move available in the tablebase, a mistake is counted if the engine suggests a move that results in a loss.

Table 1 shows that Stockfish performs better than Lc0 for the three-piece tablebases. However, for the four-piece tablebases, Lc0 shows better results than Stockfish, except for three tablebases: KQkb (win), KPkp (win), and KQkp (win). Figure 1b illustrates that Lc0 consistently outperforms Stockfish in all the tablebases, where perfect play results in a draw. Even though Stockfish shows a lower percentage of mistakes in the winning positions of KQkb, KPkp, and KQkp tablebases in Fig. 1a, Lc0 still demonstrates very competitive results in those tablebases as well. In our 1% of the total positions for different five-piece endgame tablebases, Lc0 could only perform better than Stockfish in KRBkp (draw), KBNkp (win), and KBNkp (draw). Calculating from Table 1, we find that Stockfish makes an overall 1.47% and 1.67% of errors in winning and drawing positions, respectively, whereas Lc0 makes 1.32% and 1.07% of errors.

Based on this result, we conduct a study on the Average Centipawn Loss (ACPL) to gain further insights. A Centipawn[4] is 1/100 of a pawn used to evaluate a chess position. ACPL identifies how much 'value' a player loses while playing a wrong move. An ACPL close to zero indicates a very strong move. We choose to calculate the ACPL for the 4-piece tablebases of our datasets, as the results on the three-piece tablebases are almost perfect, and we have limited data on the five-piece tablebases.

To calculate the ACPL, we divide the mistaken positions of each 4-piece tablebase into two parts: (a) Positions where Stockfish plays the correct move but Lc0 plays the wrong move, and (b) Positions where Lc0 plays the correct move but Stockfish plays the wrong move. We choose this division to assess the relative strength of the incorrect move compared to the correct one. In (a), we obtain an ACPL of -408.28 with a standard deviation of 1739.31. In (b), we obtain an ACPL of 281.71 with a standard deviation of 1476.47. These values portray that both engines recognize their moves as very weak. However, when Lc0 plays the wrong move, the ACPL is further from zero compared to the other case. This suggests that Lc0 evaluates its mistaken position more accurately than Stockfish in this study.

---

[4] https://chess.fandom.com/wiki/Centipawn.

Table 1. Total number of mistakes by the policy net of Stockfish and Lc0.

#Pieces	EGTB	W/D	Total Positions	Stockfish(Policy)	Lc0(Policy)
3	KQk	W	18081	19 **(0.1%)**	173 (0.96%)
		D	2896	0	0
	KBk	D	52234	0	0
	KNk	D	53806	0	0
	KRk	W	21959	0 **(0%)**	23 (0.1%)
		D	2796	0	0
4	KQkb	W	701738	787 **(0.11%)**	2638 (0.38%)
		D	220956	843 (0.38%)	538 **(0.24%)**
	KQkq	W	934428	18038 (1.93%)	14682 **(1.57%)**
		D	1293823	8874 (0.7%)	6714 **(0.52%)**
	KQkr	W	890800	8512 (0.96%)	7057 **(0.8%)**
		D	49184	4745 (9.6%)	2261 **(4.6%)**
	KRkr	W	784918	12759 (1.63%)	1839 **(0.23%)**
		D	1892778	16313 (0.9%)	7311 **(0.4%)**
	KPkp	W	321303	5130 **(1.6%)**	5857 (1.82%)
		D	248509	8411 (3.38%)	3580 **(1.44%)**
	KRkp	W	1110806	39490 (3.56%)	13028 **(1.17%)**
		D	398282	16417 (4.12%)	12508 **(3.14%)**
	KQkp	W	945359	4359 **(0.46%)**	7762 (0.82%)
		D	155352	2284 (1.47%)	1461 **(0.94%)**
5	KQBkp	W	1050708	4215 **(0.4%)**	12011 (1.14%)
		D	252429	215 **(0.08%)**	1560 (0.62%)
	KQNkp	W	1148101	7136 **(0.62%)**	11923 (1.04%)
		D	265648	513 **(0.19%)**	2085 (0.78%)
	KQRkp	W	931942	2388 **(0.26%)**	10634 (1.14%)
		D	30495	165 **(0.54%)**	1049 (3.44%)
	KRBkp	W	1274054	11684 **(0.92%)**	16498 (1.29%)
		D	339055	4566 (1.35%)	3924 **(1.16%)**
	KRNkp	W	1081372	16086 **(1.49%)**	19198 (1.78%)
		D	282217	5000 **(1.77%)**	6178 (2.19%)
	KBNkp	W	1208553	52401 (4.34%)	40607 **(3.36%)**
		D	410661	30795 (7.5%)	14596 **(3.55%)**

5.2   Improvement Analysis After Incorporating Search

Based on Table 1, we decide to investigate the four-piece tablebases KQkr (draw), KRkp (win), KRkp (draw), and KRkr (win). This selection is motivated by

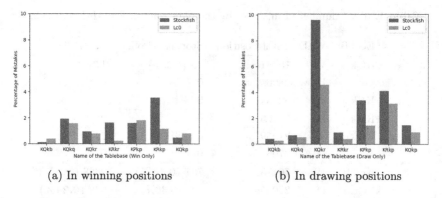

(a) In winning positions   (b) In drawing positions

Fig. 1. Performance comparison between the raw policies of Stockfish and Lc0 in the four-piece tablebases

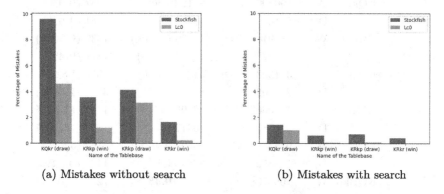

(a) Mistakes without search   (b) Mistakes with search

Fig. 2. Performance of Stockfish and Lc0 with and without search (400 Nodes)

the statistically significant performance difference between the two engines. We search up to 400 nodes with these engines to examine the changes.

Figure 2 reflects the improvement in the performance of both engines after incorporating search. In the four tablebases used here, Lc0 still performs better than Stockfish even after applying a search of up to 400 nodes. However, incorporating search leads to a significant reduction of mistakes for both engines.

5.3   Engine Behaviour Analysis in Positions with a Single Pawn for the Weaker Side

We investigate the behaviour of Stockfish and Lc0 using tablebases where there is only one black pawn and one or more stronger white pieces. We compare the behaviours of the engines based on the fact whether the black pawn is attacked or safe. Figure 3 shows examples of two boards; in one, the pawn is under attack, and in the other, the pawn is on a safe square. We consider only cases where the result is a win in perfect play. We choose to analyze this behaviour because, in these endgames, human chess players usually do not miss the chance to capture this pawn to make the opponent helpless with only the king. We decide to observe how many mistakes occur in this kind of scenario.

Table 2 displays the total number of correct moves, the percentage of attacked pawns in those positions, the total number of mistakes, and the percentage of attacked pawns in those positions for the raw networks of Stockfish and Lc0, respectively. The results show that Lc0 has a lower percentage of mistakes when the black pawn is under attack.

Interestingly, for the tablebases KQBkp, KQNkp and KQRkp, the percentages of mistakes are very high for both engines. In positions where the white can capture the black pawn, such a high rate of mistakes is not expected at all. Specifically, the performance of Stockfish for KQRkp (87.06%) is a cause for concern. Here, Stockfish makes almost 90% of its mistakes when the black pawn is under attack, whereas this percentage is 57.5% for attacked pawns in cases with no mistakes.

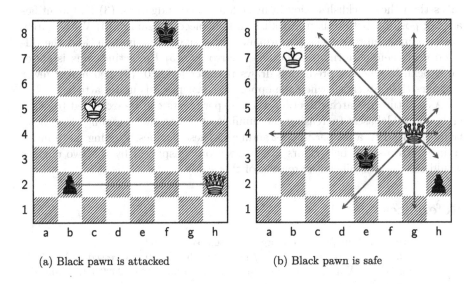

(a) Black pawn is attacked   (b) Black pawn is safe

Fig. 3. Example of black attacked and safe pawns (white to play)

Table 2. Percentages of Mistakes for Attacked and Safe Pawns (Win Only).

| | Stockfish | | | | Lc0 | | | |
| | No Mistake | | Mistake | | No Mistake | | Mistake | |
EGTB	Total Positions	% of Attacked Pawns	Total Positions	% of Attacked Pawns	Total Positions	% of Attacked Pawns	Total Positions	% of Attacked Pawns
KRkp	1071316	29.2	39490	9.95	1097774	28.82	13032	3.4
KQkp	941000	40.5	4359	35.21	937608	40.6	7751	25.21
KQBkp	1046493	52.6	4215	48.6	1038698	52.7	12010	40.84
KQNkp	1140965	48.5	7136	40.72	1136179	48.5	11922	37.61
KQRkp	929554	57.5	2388	87.06	921309	57.7	10633	50.89
KRBkp	1262371	41.3	11683	24.4	1257557	41.5	16497	19.06
KRNkp	1065286	36.5	16086	16.31	1062174	36.6	19198	13.27
KBNkp	1156153	29.07	52401	30.7	1167947	29.5	40607	18.06

6   Conclusion and Future Work

Through this work, we aim to contribute to the fascinating world of chess engines by uncovering new insights. The important findings of this work are - (1) The Stockfish policy is strictly better than or equal to the Lc0 policy in 3-piece endgames for predicting a perfect move, (2) The Lc0 policy produces fewer mistakes than the Stockfish policy in most four-piece endgames, (3) Lc0 identifies a weak position better than Stockfish in four-piece endgames, (4) With search, both engines improve their performance by a significant margin, and the difference in their performances becomes narrower, (5) Predicting wins is easier for Stockfish, whereas predicting draws is easier for Lc0, (6) Lc0 makes fewer mistakes than Stockfish when the opponent's last pawn is under attack.

As a future research direction, these experiments can be extended to other endgame tablebases. Increasing the number of pieces and sampling more positions will lead to a deeper understanding of these engines. Finding more interesting patterns and behaviours in the positions misplayed by the two engines would also be a significant extension of this study.

References

1. Lc0 authors: lc0. https://lczero.org. Accessed 04 Mar 2023
2. Lc0 authors: What is lc0? lczero.org/dev/wiki/what-is-lc0/. Accessed 04 Mar 2023
3. Lc0 authors: Win-draw-loss evaluation. lczero.org/blog/2020/04/wdl-head/. Accessed 24 Mar 2023
4. Stockfish authors: Stockfish. https://stockfishchess.org. Accessed 04 Mar 2023
5. Ballicora, M.: Gaviota. sites.google.com/site/gaviotachessengine/Home/ endgame-tablebases-1. Accessed 27 Feb 2023
6. Coulom, R.: Efficient selectivity and backup operators in Monte-Carlo tree search. In: International Conference on Computers and Games, pp. 72–83. Springer (2006)
7. Edwards, D.J., Hart, T.: The alpha-beta heuristic. Tech. Rep. AIM-30, MIT (1961)
8. Forsten, H.: Purpose of the moves left head. https://github.com/Leela ChessZero/lc0/pull/961#issuecomment-587112109. Accessed 24 Mar 2023

9. Haque, R., Wei, T.H., Müller, M.: On the road to perfection? evaluating leela chess zero against endgame tablebases. In: Advances in Computer Games: 17th International Conference, ACG 2021, Virtual Event, November 23–25, 2021, Revised Selected Papers, pp. 142–152. Springer (2022)
10. Heinz, E.A.: Extended futility pruning. ICGA J. **21**(2), 75–83 (1998)
11. Kryukov, K.: Number of unique legal positions in chess endgames (2014). http://kirillkryukov.com/chess/nulp/. Accessed 26 Feb 2023
12. Lassabe, N., Sanchez, S., Luga, H., Duthen, Y.: Genetically programmed strategies for chess endgame. In: Proceedings of the 8th Annual Conference on Genetic and Evolutionary Computation, pp. 831–838 (2006)
13. de Man, R., Guo, B.: Syzygy endgame tablebases. syzygy-tables.info/. Accessed 06 Mar 2023
14. de Man, R.: Syzygy. https://github.com/syzygy1/tb. Accessed 08 Mar 2023
15. McGrath, T., Kapishnikov, A., Tomašev, N., Pearce, A., Wattenberg, M., Hassabis, D., Kim, B., Paquet, U., Kramnik, V.: Acquisition of chess knowledge in alphazero. Proc. Natl. Acad. Sci. **119**(47), e2206625119 (2022)
16. Nasu, Y.: Efficiently updatable neural-network-based evaluation functions for computer shogi. The 28th World Computer Shogi Championship Appeal Document 185 (2018)
17. Romein, J.W., Bal, H.E.: Awari is solved. ICGA J. **25**(3), 162–165 (2002)
18. Silver, D., et al.: Mastering the game of go with deep neural networks and tree search. Nature **529**(7587), 484–489 (2016)
19. Silver, D., Hubert, T., Schrittwieser, J., Antonoglou, I., Lai, M., Guez, A., Lanctot, M., Sifre, L., Kumaran, D., Graepel, T., et al.: A general reinforcement learning algorithm that masters chess, shogi, and go through self-play. Science **362**(6419), 1140–1144 (2018)

Solving Games

Solving NoGo on Small Rectangular Boards

Haoyu Du[✉], Ting Han Wei, and Martin Müller[ID]

University of Alberta, Edmonton, Canada
{du2,tinghan,mmueller}@ualberta.ca

Abstract. The game of NoGo is similar to Go in terms of rules, but requires very different strategies. While strong heuristic computer players have been created for NoGo, solving and optimal play have been less studied. We introduce Sorted Bucket Hash (SBH), a new approach to building transposition tables for game solvers, and apply it to solve NoGo on small boards. Using boolean negamax with standard heuristics and SBH, our program SBHSolver has now solved NoGo on 50 different rectangular boards including 3×9, the largest solved NoGo game to date. It re-solved 5×5 NoGo much more efficiently than She's work in 2013 and Cazenave's work in 2020. The SBH data structure can also efficiently extract a proof tree for the game. We provide analyses of NoGo proof trees and games, and discuss human-understandable strategies from this perspective.

Keywords: Game solving · Transposition tables · NoGo · Sorted Bucket Hash

1 Introduction

NoGo (or *Anti-Atari Go* [5]) is a lesser-known variant of the widely studied game of Go. The rules of NoGo are:

1. Black and White take turns to play, with Black going first. At each turn a stone of the player's colour is placed onto an empty point on the board. Passing is forbidden.
2. Connected stones of the same colour form a *block*. Adjacent empty points of blocks are called *liberties*.
3. All blocks must always have at least one liberty. For Go players, this means that both suicide and capturing are forbidden.
4. The game ends when a player has no legal move to play. This player is deemed the loser.

The simple twist to the rules makes NoGo very different to play. Unlike Go, where *ko* fights can greatly extend the length of a game, the number of moves in NoGo is strictly less than the size of the board, since each block needs an empty point for a liberty. The game state is completely determined by the current

© The Author(s), under exclusive license to Springer Nature Switzerland AG 2024
M. Hartisch et al. (Eds.): ACG 2023, LNCS 14528, pp. 39–49, 2024.
https://doi.org/10.1007/978-3-031-54968-7_4

board. However, NoGo still has a state space that grows exponentially with the number of points on the board, and no easy winning strategies are known.

The three main contributions in this paper are:

1. The data structure of Sorted Bucket Hash (SBH) for space-efficient transposition tables based on perfect hashing. SBH is designed for use in weakly solving games.
2. Efficient solutions of NoGo on all 50 rectangular boards of size up to 27 points, with orders of magnitude fewer nodes than the 5 × 5 NoGo solution by She [9].
3. New results on NoGo, such as a winning strategy for Black on 5 × 5 in at most 21 moves, two general results for 1 × n NoGo, and a statistical analysis of two human-understandable heuristics from the solver's perspective.

2   Related Work

NoGo is a relatively young game with few human players. Previous research mostly focused on creating strong computer agents for competitions such as the Computer Olympiad, TAAI (Taiwanese Association for Artificial Intelligence), and TCGA (Taiwan Computer Game Association) tournaments. Early programs include BobNoGo [4], an open-source program based on MCTS that includes an exact solver.

NoGo was proven a win for the first player (Black) on the 3 × 3 board and a loss on 4 × 4 in 2011 [6]. All three distinct opening moves on 3 × 3 boards win. 5 × 5 NoGo was solved in 2013 by Pohsuan She [9]. Black can win with all six distinct opening moves, as shown in Fig. 1. Cazenave determined the winner of NoGo played on boards of size up to 25 points by using alpha-beta search with Monte Carlo Move Ordering and a transposition table of size 1048575 entries [1].

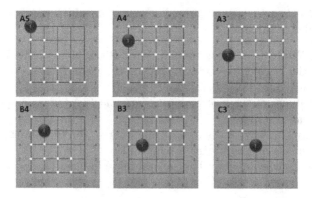

Fig. 1. The six distinct openings for 5 × 5 NoGo, from [9]. White dots represent distinct possible replies from White.

3 Sorted Bucket Hash and Its Use in Solving Games

3.1 Boolean Negamax Search and Heuristics

While Sorted Bucket Hash is a general data structure, we explore it in the specific context of transposition tables for Boolean Negamax, which is a simpler special case of alpha-beta search [3] for two-valued outcomes. Our search also uses two standard heuristics, Enhanced Transposition Cutoff [7] and History Heuristic [8].

3.2 A Transposition Table with Perfect Hashing

A transposition table is useful to avoid redundant evaluation when an equivalent game position can be reached via different move sequences. In typical implementations, each game state is mapped to a k-bit hash code. An m-bit part of the code is used as an index into a hash table, and the remaining $n = k - m$ bits are used as a validation code which is stored with each hash entry.

Tradeoffs for Designing Hash Tables. Designing an efficient hash table requires navigating trade-offs between speed, memory used, and the amount of information stored per position. Our solver is designed for the following scenario:

- Large state space, with many transpositions, so maximizing the size of the table is important.
- Storing all solved positions, in contrast to using a fixed memory table with a replacement scheme and re-search [2].
- The amount of data per position that needs to be stored is minimal. One bit for storing win/loss is enough (a few extra bits are useful as discussed below).
- Perfect hashing, as discussed below.

Perfect Vs Lossy Hashing With perfect hashing, each game position is mapped to a unique hash code. For example, all the states of a NoGo board with n points can be mapped to 3^n distinct codes. If storing a full hash code takes too much space, lossy hash functions such as 64-bit Zobrist Hashing [10] are used in practice. However, in our application 64-bit Zobrist codes used too much memory, and smaller codes led to too many hash collisions. This was the main motivation for developing SBH based on perfect hashing.

Sorted Bucket Hash Sorted Bucket Hash (SBH) is a new method for organizing a transposition table. Given a game state s, SBH uses a perfect hash function $h(s)$ that produces $k = m + n$-bit hash codes. The m-bit segment of a hash code represents the bucket index, and the n-bit segment serves as the validation code. A SBH hash table consists of 2^m buckets. Each bucket holds at most 2^n entries. Buckets are empty at the beginning and become populated as code-value pairs are stored. A bucket entry consists of a 1-bit game value and an n-bit validation code. SBH keeps the entries in each bucket sorted by their validation codes. Binary search based on validation codes is used to find entries and insertion points in a bucket. The find operation of SBH takes $O(1)$ for the hashing part, and $O(n)$ for binary search among the at most 2^n validation codes inside a bucket. This compares favorably with the $O(2^n)$ linear search in chaining.

The values of k, m, and n are problem-dependent and should be carefully selected by finding the balance between algorithm efficiency and the actual memory size. Below, we discuss our choices for solving NoGo.

SBH Find and Store Operations. The operations find and store are implemented as follows: A given k-bit hash key is split into m-bit index and n-bit validation code. The index selects the bucket, and binary search of the validation code completes the find operation. Store involves a find of the correct location within the right bucket, followed by allocating a new one larger array and copying the old data over in two parts, with the new entry stored in between.

Collecting Solutions in SBH. SBH provides an easy and efficient way to extract a winning strategy from the transposition table after a successful search.

This uses an extra "proof flag" and an encoded winning move stored in each hash entry (see details for NoGo below). Solution extraction marks all nodes that are part of the proof tree, starting with the root. Another Boolean Negamax "search" is guided by the results stored in the transposition table: at each OR node, the stored move is chosen to find a child node, while in an AND node all children are traversed. The proof flag is set for all the nodes encountered, and the set of marked nodes forms the solution.

To actually store the solution to disk, sequentially visit all buckets and write out all marked nodes. The size of the stored solution is typically much smaller than the original transposition table after the search. This stored solution is also sorted by full hash codes, making it easy to reload the solution into the SBH transposition table for game playing or analysis later. For example, storing the solution takes less than a minute for 5×5 NoGo on modest hardware.

SBH Implementation for NoGo. We discuss implementation details of SBH for our program SBHSolver in the case of 5×5 NoGo, where a comparison to previous work is possible. For hashing, a board is encoded as a 25-digit base 3 number, with point encoding empty $= 0$, black $= 1$, and white $= 2$. Since $2^{39} < 3^{25} < 2^{40}$ this requires using $k = 40$-bit hash codes. A small m results in poor performance due to many positions being hashed into the same bucket, while a large m potentially leads to more memory overhead. Based on our hardware (32GB of memory) and some experimentation, we used an $m = 30$-bit index and an $n = 40 - 30 = 10$-bit validation code, with a bucket size limit of $2^{10} = 1024$. An example of calculating the hash code, index, and validation code of a NoGo board is shown in Fig. 2.

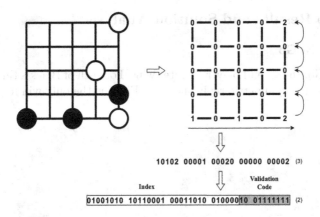

Fig. 2. The calculation of hash code, index, and validation code of a NoGo board.

5×5 NoGo has over 8.47×10^{11} possible states that need to be addressed, requiring 40 bits. A typical proof of one opening move visited about $2.42 \times 10^8 < 2^{28}$ distinct game positions, less than 0.03% of the represented space of 2^{40}.

SBHSolver is implemented in C++. The hash table is an array of 2^{30} pointers as shown in Fig. 3. Each pointer points to a bucket, or is null if the bucket is empty. A bucket is a dynamically allocated sorted array with at most 2^{10} entries. Each entry occupies 2 bytes and consists of 1-bit proof flag, 5-bit winning move/ outcome, and 10-bit validation code. The 5-bit winning move represents either a winning legal move for the current player or an illegal move encoded as 11111. The illegal move 11111 implies a loss, while any legal move implies a winning state. The proof flag is used for collecting a solution. Bucket entries are sorted by validation code in ascending order. The size of a bucket grows by 1 with each insertion.

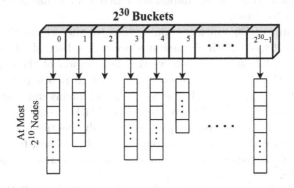

Fig. 3. Data structure for 5×5 NoGo using Sorted Bucket Hash with $m = 30$, $n = 10$. Each cube represents one of the 2^{30} buckets. Each bucket stores up to 2^{10} positions sorted by validation code. In this example, bucket 2 is empty.

4   NoGo Results and Solution Analysis

4.1   4 × 4 NoGo

In 4 × 4 NoGo, the second player (White) wins. For each of Black's three distinct initial moves, the symmetric reply shown in Fig. 4 is the only win for White.

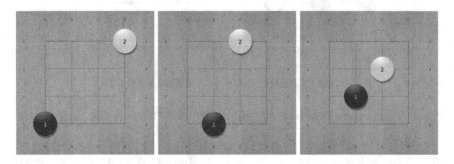

Fig. 4. The 3 distinct Black opening moves and White's unique winning replies.

Playing symmetrically does not win in all 4 × 4 positions where it applies, but it is an effective heuristic. Among 24,708 game positions where a white move can make the board symmetric, such a move wins 85.3% of the time.

4.2   5 × 5 NoGo

SBHSolver proved that Black wins with all six distinct opening moves shown in Fig. 1, confirming the result from She [9]. As a new result, we proved that Black can force a win in at most 21 moves for all six openings, but cannot force a win in 19 moves or less. We showed these by introducing a threshold $T = 21$ (or $T = 19$), and modifying the evaluation such that games that are not won in T moves are losses for Black.

As an example of the efficiency gains, to solve the A1 opening SBHSolver evaluated 2,968,264,746 distinct game positions. A subset of 242,002,061 positions forms the solution. Fig. 5a shows the distribution of these proven game positions by depth. The number of nodes is largest at depth 18, then decreases due to two factors: more games being decided, and having more transpositions.

There are fewer white-to-play positions than black-to-play positions one ply earlier, since for each white-to-play position we only identify one winning black move, and there are transpositions among the resulting black-to-play positions. The frequency of transpositions within the solution increases with depth, as shown in Fig. 5b.

SBHSolver is efficient compared to She's solution [9]: using SBH with boolean minimax and two standard heuristics, SBHSolver searches on average 188× fewer nodes in solving the six distinct openings. It is also 15× faster than Cazenave's solution [1] that evaluated 46,092,056,485 moves.

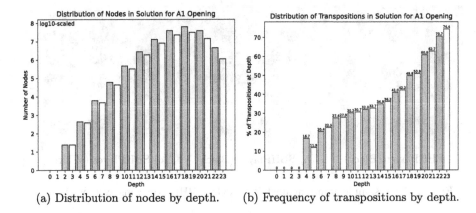

(a) Distribution of nodes by depth.   (b) Frequency of transpositions by depth.

Fig. 5. Analysis of a solution to 5x5 NoGo with the A1 opening, by search depth.

Sample 5 × 5 Games. Our perfect SBHPlayer follows one of our winning strategies for Black. We tested it against the strong open source program BobNoGo [4]. Figures 6a and 6b show two representative wins. The first game follows the original A1 proof, and the second game the 21-move solution of C3.

In the third game in Fig. 6c SBHPlayer wins against BobNoGo even as White, from the losing side. At move three, SBHPlayer knows that E2 is winning for Black. However, BobNoGo chose C3, and SBHPlayer quickly exploited this weak move and solved this variation by search, partially relying on the precomputed solution to avoid mistakes, and won in 22 moves.

Comments on NoGo Strategy. Making eyes, an important element of Go strategy, is often recommended for NoGo as well, and is frequently seen from the MCTS-based BobNoGo player. Eyes represent a local advantage, "reserving" points for the player. However, only 45.2% of the terminal positions in the A1 solution contain any eyes for Black, so SBHPlayer very often wins without them. Instead, SBHPlayer seems to favor creating long blocks that separate the opponent's stones. This seems to be an effective strategy, at least in 5 × 5 NoGo. Both behaviors are seen in all the games in Fig. 6: SBHPlayer creates long blocks, but no eyes.

4.3   Solutions of Rectangular NoGo Boards

Figures 7 and 8 summarize the results of SBHSolver on solving rectangular NoGo boards. It solved all such boards with up to 27 empty points. Black wins on most boards. Notable exceptions, which are White (second player) wins, are 1×1, $4 \times n$ with $n \leq 4$, and several $2 \times n$ boards. The results on 3×9, $1 \times n$ with $10 < n \leq 27$, and $2 \times n$ with $10 < n \leq 13$ are new compared to [1].

$1 \times n$ **NoGo.** Figure 8 summarizes the win/loss results of all first moves on $1 \times n$ NoGo boards for $n \leq 27$. A black stone indicates a winning move for

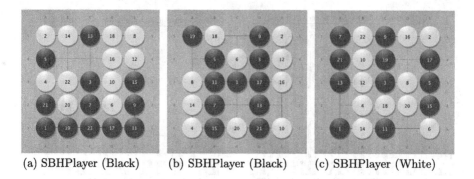

(a) SBHPlayer (Black)   (b) SBHPlayer (Black)   (c) SBHPlayer (White)

Fig. 6. Three sample games against BobNoGo. SBHPlayer wins even as White.

Column

Row	1	2	3	4	5	6	7	8	9	10	11	12	13
1	0	1	1	0	1	1	1	1	1	1	1	1	1
2		1	0	0	1	1	1	1	0	0	1	1	1
3			1	0	1	1	1	1	1				
4				0	1	1							
5					1								

Fig. 7. Table of game theoretic values for rectangular NoGo boards. 1 means a win for Black from the empty board, 0 a loss. Symmetric results for *row* > *column* are omitted to save space. For more $1 \times n$ results see Fig. 8.

Black, a white stone a loss. Black wins with most opening moves, especially on large boards. Symmetry arguments lead to two general results for arbitrary n:

Theorem 1. *In $1 \times n$ NoGo, with odd $n > 1$, the center point $(n+1)/2$ wins.*

Proof. The Black move at $(n+1)/2$ divides the board into two subgames. A winning strategy for Black is to mirror the previous White move. Whenever a white move is legal because of a liberty at some point x, the mirror black move is also legal because of a corresponding liberty at $n + 1 - x$. An example of this strategy is shown in Fig. 9.

Theorem 2. *In $1 \times n$ NoGo, with even $n > 2$, the two middle points $n/2$ and $n/2 + 1$ lose.*

Proof. If Black plays at one middle point, a winning strategy for White is to play the other, then follow the mirroring strategy. The game splits into two independent subgames G and $-G$ in terms of Combinatorial Game Theory, and

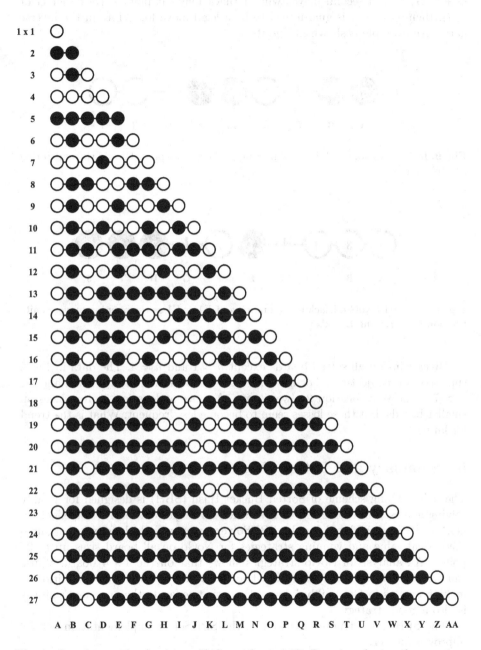

Fig. 8. Opening results for $1 \times n$ NoGo with $n \leq 27$. Row i evaluates the opening moves of $1 \times i$ NoGo (black = winning move, white = losing). For example, in 1×7 NoGo, the only winning first move for Black is D1 in the middle.

$G + (-G) = 0$, a second player win. If Black plays at point x (in either G or $-G$), then $n + 1 - x$ is guaranteed to be a legal move for White in the inverse game. An example is shown in Fig. 10.

Fig. 9. In 1×9 NoGo, Black wins by playing at the middle point E1 and then mirroring White's moves.

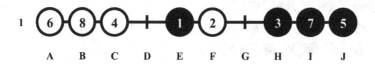

Fig. 10. In 1×10 NoGo, Black loses if playing at E1 or F1 as the opening move. White can win by mirroring Black's moves.

Black wins on all solved boards except $n = 1$ and $n = 4$. The data in Fig. 8 supports two conjectures: For $n > 5$, a move at location 1 always loses, and for $n > 7$, a move at location 2 always wins. A move at location 3 often loses on smaller boards, but these losses seem to become less frequent. What is the trend for larger n?

5   Summary and Future Work

The new hashing scheme of Sorted Bucket Hash (SBH) is designed for weakly solving games, and extracting their solution strategies efficiently. The effectiveness of SBH is validated by the SBHSolver program, which found the game-theoretical value and winning strategies for NoGo on all board sizes up to 27 points. An analysis of NoGo strategies shows that on the 4×4 board, White can often benefit from playing symmetrically. On the 5×5 board, in addition to making eyes as in Go, building long blocks that separate the opponent's stones is also a good strategy.

The SBH data structure as well as the SBHSolver implementation can be improved further:

1. With the current simple encoding, many game positions hash into the same buckets, slowing down find and store operations over time. Spreading out the perfect hash codes, such as with a linear congruence mapping, should give better distribution across buckets, potentially improving the performance.

2. Depending on game encoding details, large stretches of consecutive buckets may remain empty. A better data structure could compress long stretches of null pointers.
3. SBHSolver reallocates bucket memory at each insertion. An alternative strategy such as doubling the bucket size would reduce memory copies at the cost of larger storage requirements.
4. To solve even larger games, where the search does not fit into main memory, SBH should be combined with a two-tier (disk+memory) storage scheme.

References

1. Cazenave, T.: Monte carlo game solver. In: Monte Carlo Search, MCS 2020. Communications in Computer and Information Science, vol. 1379, pp. 56–70 (2021)
2. Kishimoto, A.: Correct and Efficient Search Algorithms in the Presence of Repetitions. Ph.D. thesis, University of Alberta (2005)
3. Knuth, D.E., Moore, R.W.: An analysis of alpha-beta pruning. Artif. Intell. 6(4), 293–326 (1975)
4. Müller, M.: The BobNoGo program. https://webdocs.cs.ualberta.ca/~mmueller/nogo/BobNoGo.html. Accessed 31 May 2023
5. Müller, M.: NoGo history and competitions. https://webdocs.cs.ualberta.ca/~mmueller/nogo/history.html. Accessed 31 May 2023
6. Müller, M.: Solving NoGo on small board sizes. https://webdocs.cs.ualberta.ca/~mmueller/nogo/solving.html. Accessed 31 May 2023
7. Plaat, A., Schaeffer, J., Pijls, W., de Bruin, A.: Exploiting graph properties of game trees. In: AAAI/IAAI, vol. 1 (1996)
8. Schaeffer, J.: The history heuristic. ICGA J. 6(3), 16–19 (1983)
9. She, P.: The Design and Study of NoGo Program. Master's thesis, National Chiao Tung University (2013)
10. Zobrist, A.L.: A new hashing method with application for game playing. ICCA J. 13(2), 69–73 (1990)

Optimal Play of the Great Rolled Ones Game

Todd W. Neller[⊠], Quan H. Nguyen, Phong T. Pham, Linh T. Phan, and Clifton G.M. Presser

Gettysburg College Department of Computer Science, Gettysburg, USA
`tneller@gettysburg.edu`

Abstract. In this paper, we solve and visualize optimal play for the Great Rolled Ones jeopardy dice game by Mitschke and Scheunemann [4, p. 4–5]. We share the second player advantage and compute that the first player should start with 3 compensation points (komi) for greatest fairness. We present a spectrum of human-playable strategies that trade off greater play complexity for better performance, and collectively clarify key considerations for excellent play.

1 Introduction

Great Rolled Ones is a jeopardy dice game first published in 2020 by Sam Mitschke and Randy Scheunemann [4, p. 4–5] that is similar to the game Zombie Dice [1]. Both are jeopardy dice games [3, Ch. 6] in the Ten Thousand dice game family [2]. In this paper, we analyze Great Rolled Ones, computing optimal play as well as providing additional insights to gameplay.

We begin by describing the rules of Great Rolled Ones, and then define 2-player optimality equations and our method for solving them. We calculate compensation points (komi) for a fairest game, visualize the policy, and share observations on the optimal roll/hold boundary. We then present an array of human-playable policies we have devised along with their performances against the optimal policy. The policies demonstrate different design trade-offs of greater complexity for greater win rates, and highlight key play policy considerations. Finally, we discuss future work and summarize our conclusions.

2 Rules

Great Rolled Ones (GRO) is a dice game for two or more players using 5 standard (d6) dice. In this paper, we will focus on the two-player GRO game. Players will have the same number of turns. A *turn* consists of a sequence of player dice rolls where rolled 1 s are set aside. The turn ends when either the player decides to *hold* (i.e. stop rolling) and score the total number of non-1 s rolled, or has rolled three or more 1 s, ending the turn and scoring 0 points. A *round* consists of each player taking one turn in sequence. Any player ending their turn with a goal

© The Author(s), under exclusive license to Springer Nature Switzerland AG 2024
M. Hartisch et al. (Eds.): ACG 2023, LNCS 14528, pp. 50–59, 2024.
https://doi.org/10.1007/978-3-031-54968-7_5

score of 50 or more causes that to be the last round of the game. At the end of the last round, the player with the highest score wins.

Given that the rules refer to a singular winner ("cultist with the most rituals", i.e. player with the most points) after the last round where "everyone else loses", this implies that no player can opt to draw, and thus a player is constrained to attempt to exceed the score of the current leader in the last round. An optimal player must attempt to win when a prior player of that round has reached 50 or more points.

Example round:

- Player 1 initially rolls $\{1, 1, 3, 4, 5\}$. Two 1 s were rolled and set aside, so 3 is added to the *turn total* of non-1 s rolled. Player 1 chooses to *roll* the remaining three non-1 dice again with a result of $\{2, 2, 4\}$. No 1 s were rolled and set aside, so 3 is again added to the turn total for a new turn total of 6. Player 1 chooses to roll the three remaining non-1 dice again with a result of $\{1, 1, 6\}$. Two 1 s are set aside, for a total of four 1 s. Three or more 1 s ends a turn scoring 0 points, so play passes to player 2 with no score change.
- Player 2 initially rolls $\{4, 4, 4, 4, 5\}$, sets no 1 s aside, has a turn total of 5, and chooses to roll again. Player 2 rolls $\{4, 4, 4, 5, 5\}$, setting no 1 s aside, has a new turn total of 10, and chooses to roll again. Player 2 rolls $\{1, 1, 2, 4, 5\}$, sets two 1 s aside, has a new turn total of 13, and chooses to *hold*, scoring 13 points and ending the round.

The game thus consists of roll/hold risk assessment in a race to achieve the top score of 50 or more points within the same number of turns as other players. Given the player scores, the turn total, and the number of 1 s set aside, should the current player roll or hold so as to maximize the probability of winning?

3   Optimality Equations and Solution Method

We here define optimality equations for the GRO two-player game where player 2 must seek to exceed player 1's score when it is at least 50.

Nonterminal states are described as the 5-tuple (p, i, j, k, o), where p is the current player number (1 or 2), i is the current player score, j is the opponent score, k is the turn total, and o is the number of rolled 1 s set aside.

Let $P_{\text{new1s}}(d, o_{\text{new}})$ denote the probability that o_{new} of d dice rolled are 1 s $(0 \leq o_{\text{new}} \leq d \leq 5)$:

$$P_{\text{new1s}}(d, o_{\text{new}}) = \binom{d}{o_{\text{new}}} \left(\frac{1}{6}\right)^{o_{\text{new}}} \left(\frac{5}{6}\right)^{(d - o_{\text{new}})}$$

Let $P_{\text{exceed}}(\Delta, o)$ denote the probability that player 2 will exceed player 1's score ≥ 50 where $\Delta = j - (i + k)$ (their score difference) and o is the number of rolled 1 s set aside on player 2's final turn. Then,

$$P_{\text{exceed}}(\Delta, o) = \begin{cases} 0 & \text{if } o \geq 3 \\ 1 & \text{if } \Delta < 0 \\ \sum_{n=0}^{2-o} P_{\text{new1s}}(5 - o, n) P_{\text{exceed}}(\Delta - (5 - o'), o') & \text{otherwise} \\ \quad \text{where } o' = o + n \end{cases}$$

The probability of winning with a roll $P_{\text{roll}}(p, i, j, k, o)$ under the assumption of optimal play thereafter is:

$$P_{\text{roll}}(p, i, j, k, o) = \begin{cases} P_{\text{exceed}}(j - i, o) & \text{if } p = 2 \\ & \text{and} \\ & j \geq 50 \\ \sum_{n=0}^{2-o} P_{\text{new1s}}(5 - o, n) P(p, i, j, k + 5 - o', o') + & \\ \sum_{n=3-o}^{5-o} P_{\text{new1s}}(5 - o, n)(1 - P(3 - p, j, i, 0, 0)) & \text{otherwise} \end{cases}$$

A player can (and should) never hold at the beginning of the turn when the turn total is 0, so we express this by treating such rule-breaking as a loss. Thus, the probability of winning with a hold $P_{\text{hold}}(p, i, j, k, o)$ under the assumption of optimal play thereafter is:

$$P_{\text{hold}}(p, i, j, k, o) = \begin{cases} 0 & \text{if } k = 0 \text{ or } (p = 2 \text{ and } j \geq 50, i) \\ 1 & \text{if } p = 2 \text{ and } i + k \geq 50, j \\ 1 - P(3 - p, j, i + k, 0, 0) & \text{otherwise} \end{cases}$$

Then the probability of winning $P(p, i, j, k, o)$ under the assumption of optimal play is:

$$P(p, i, j, k, o) = \max(P_{\text{roll}}(p, i, j, k, o), P_{\text{hold}}(p, i, j, k, o))$$

What remains is to bound our nonterminal states for computation. The rules have no restriction on how high a turn total (and thus a score) can go. Our approach is to create a high enough artificial maximum score M (e.g. 100) bounding i, j, and k such that optimal policy does not expand play to further nonterminal states for any tested increase in M.

Having bounded our nonterminal state space representation such that $p \in \{1, 2\}, 0 \leq i, j, k \leq M, 0 \leq o \leq 2$, we apply value iteration as in [5] until the maximum probability change of an iteration is less than $\epsilon = 10^{-14}$.

4   Optimal Policy

The optimal roll/hold boundaries of GRO are shown in Fig. 1. Each subfigure depicts a 3-dimensional (i, j, k) roll/hold boundary for each possible pair of player p and rolled ones o. Axes are player score i, opponent score j, and turn total k. Given a current state inside or outside of the appropriate solid, an optimal player should roll or hold, respectively.

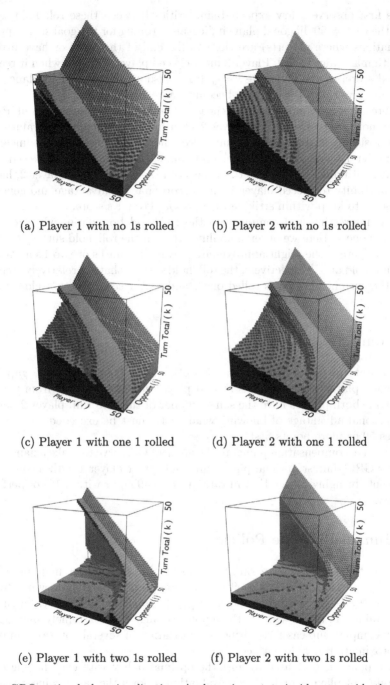

(a) Player 1 with no 1s rolled   (b) Player 2 with no 1s rolled

(c) Player 1 with one 1 rolled   (d) Player 2 with one 1 rolled

(e) Player 1 with two 1s rolled   (f) Player 2 with two 1s rolled

Fig. 1. GRO optimal play visualization. A player in a state inside or outside the gray solid should roll or hold, respectively. Subfigures are by p, o cases, and axes follow i, j, k state variables. (Color figure online)

We first observe a few expected similarities between these roll/hold solids. First, the $i + k = 50$ diagonal plane indicating a rolling for the goal score appears in situations where player(s) are close to the end of the game or have little to risk with many dice to roll. Player 2 must exceed player 1's score when it reaches the goal score, so the plane $i + k = j + 1$ is also a prominent hold plane. As a player has fewer dice to roll, play becomes more conservative.

There are some interesting differences and subtleties to observe as well. Player 1 plays more aggressively than player 2 with higher minimum hold values with all other state variables being equal. Also, there are interesting nonlinearities when player 1 seeks to not just reach 50 points, but to far enough exceed 50 so as to make it unlikely that player 2 will exceed their final score. Player 2, having the opportunity to exceed player 1's final score, has an advantage and generally plays so as to keep within striking distance of player 1's score.

Most interesting and complex are the roll/hold boundaries when a player has rolled one 1. Here we observe nonlinearities in the roll/hold surface for both players. Whereas one might approximate player with no 1 s or two 1 s as "always roll" and "hold at 5", respectively, the roll/hold surface shape is relatively complex when the current player has rolled one 1 and player scores are not close to the goal.

5   Komi

The win rate of player 1 when play is optimal is ~ 0.4495, a $\sim 10\%$ gap from the second player win rate. This second player advantage comes from the fact that, while both players have the same number of turns to win, player 2 has the informational advantage of knowing what score must be exceeded when player 1 scores 50 or more points first.

Komi, i.e. compensation points in the game of Go, serve to make a game more fair. For GRO, fairest optimal play komi would start player 1 with 3 compensation points bringing player 1's win rate up to 0.4955, or within 1% of perfectly fair play.

6   Human-Playable Policies

In this section, we present a range of human-playable policies mapping states to roll/hold actions that trade off greater complexity for greater win rate. By *human-playable*, we mean that all roll/hold decisions may be made through simple mental math. As we will see, these policies range from extremely simple rules to very-complex sub-cases requiring memorization of several constants in order to approximate roll/hold surfaces.

Each policy is evaluated against the optimal policy with each having equal probability of playing first. Policy evaluation follows the same value-iteration-style algorithm of [6]. The performance of each is summarized in Fig. 2.

We present each policy as a method that returns whether or not to roll in the given state.

Policy	Difference
Roll with 4 or 5 Dice	-0.0536
Fixed Hold-At	-0.0268
Simple Player and Ones Cases	-0.0201
Keep Pace, End Race, by Case	-0.0100

Fig. 2. Differences between human-playable and optimal policy win rates

6.1   Roll with 4 or 5 Dice Policy

Simplest is to always roll 4 or 5 dice (unless player 2 can hold and win), and always hold with 3 or fewer dice (unless player 2 must exceed player 1's game-ending score):

Algorithm 1: Roll with 4 or 5 dice

Input   : player p, player score i, opponent score j, turn total k, ones rolled o
Output: whether or not to roll
1 **if** $p = 2 \land j \geq 50 \land i + k \leq j$ **then**   // player 2 must exceed player 1
2 | **return** *true*
3 **else if** $p = 2 \land i + k \geq 50$ **then**   // player 2 must hold at goal score
4 | **return** *false*
5 **else**   // roll with 4 or 5 dice
6 | **return** $o < 2$
7 **end if**

Surprisingly, Algorithm 1 wins only ~5.4% less than the optimal policy. For all of the nuances of optimal play, this trivial baseline performance immediately hints at high human play performance possibilities.

6.2   Fixed Hold-At Policy

Next, we consider a policy where we need only remember a few turn total thresholds.

Requiring memorization of only two hold-at constants (24 and 4), Algorithm 2 reduces the optimal play gap to ~2.7%.

6.3   Simple Player and Ones Cases

The fixed hold-at policy had the same play policy for both players with the exception of player 2's game-ending constraints. This next policy breaks down cases not only by number of ones rolled o, but also by current player number p.

Algorithm 3 also requires memorization of only two constants (20 and 5), yet requires more case memorization. Even so, breaking down cases according to

---

Algorithm 2: Fixed hold-at

Input   : player p, player score i, opponent score j, turn total k, ones rolled o
Output: whether or not to roll

1 **if** $p = 2 \wedge j \geq 50 \wedge i + k \leq j$ **then**   // player 2 must exceed player 1
2 | **return** *true*
3 **else if** $i + k \geq 50$ **then**   // player 2 holds and wins
4 | **return** *false*
5 **else if** $o = 0$ **then**   // keep rolling with 5 dice
6 | **return** *true*
7 **else if** $o = 1$ **then**   // hold at 24 with 4 dice
8 | **return** $k < 24$
9 **else**   // hold at 4 with 3 dice
10 | **return** $k < 4$
11 **end if**

Algorithm 3: Simple player and ones cases

Input   : player p, player score i, opponent score j, turn total k, ones rolled o
Output: whether or not to roll

1 **if** $p = 1$ **then**   // player 1 cases
2 | **if** $o = 0$ **then**   // keep rolling with 5 dice
3 | | **return** *true*
4 | **else if** $o = 1$ **then**   // hold at goal with ≥ 20 lead with 4 dice
5 | | **return** $k < \max(50 - i, 20 + j - i)$
6 | **else**   // hold at 5 or goal with 3 dice
7 | | **return** $k < \min(50 - i, 5)$
8 | **end if**
9 **else**   // player 2 cases
10 | **if** $j \geq 50$ **then**   // player 2 must exceed player 1
11 | | **return** $i + k \leq j$
12 | **else if** $i + k \geq 50$ **then**   // hold at goal score
13 | | **return** *false*
14 | **else if** $o < 2$ **then**   // roll with 4 or 5 dice
15 | | **return** *true*
16 | **else**   // hold at 5 or goal with 3 dice
17 | | **return** $k < \min(50 - i, 5)$
18 | **end if**
19 **end if**

---

player p and the number of ones rolled o impressively reduces the optimal play gap to ~2.0%.

In the trade-off of increased cognitive complexity for increased performance, this algorithm might represent a preferred middle ground for players. In prose, we might describe this policy as follows:

> For player 1, roll with 5 dice. With 4 dice, hold at or beyond the goal with a lead of at least 20. For player 2, if player 1 has reached the goal score, exceed it. Otherwise, if player 2 can hold and win, do so. Otherwise, player 2 always keeps rolling with 4 or 5 dice. With 3 dice, both players should hold if it reaches the goal score or if the turn total is at least 5.

6.4   Keep Pace, End Race, by Case

Algorithm 4 also breaks down cases by player p and number of ones set aside o, computing hold-at values sensitive to score difference $\delta = j - i$ combined with roll-to-the-end thresholds.

This policy requires even more case analysis, being sensitive to individual scores or score sums reaching progress thresholds. Ten constants are considerably more to remember, as well. Still, this extra work even better approximates optimal play performance, closing the optimal play gap to ~1.0%.

7   Future Work

One might use supervised learning on our roll/hold or win probability tables to compress a good play policy in memory without significantly sacrificing performance. Given that relatively simple human playable policies can perform within a few percent of optimal, it would be interesting to see what memory reductions via supervised learning are possible that closely approximate optimal play.

We conjecture that memory-efficient supervised learning models that approximate the probability of winning for each (p, o) pair could be used with a one-step backup of optimality equations in order to make an excellent, compact computational approximation of optimal play policy.

Another possibility for future work is to survey expected game lengths of the most popular jeopardy dice games, and tune the GRO goal score so as to optimize its game length. In games of chance, there is a trade-off between game brevity and the rewarding of player skill. With few decisions, a player's skill is difficult to discern through the game's variance. With many decisions, a player's skill will be rewarded with a noticeable gain in win rate (e.g. backgammon). However, a game of chance with too many decisions can become tedious.

We believe there is potential to use our analytical tools or reinforcement learning approximations of optimal play in order to advance AI-assisted game design for jeopardy dice games.

Algorithm 4: Keep pace, end race, by case

Input : player p, player score i, opponent score j, turn total k, ones rolled o
Output: whether or not to roll

1   $\delta \leftarrow j - i$
2   **if** $p = 1$ **then**   // player 1 cases
3   **if** $o = 0$ **then**   // hold at goal with $\geq$ 38 lead with 5 dice
4 | **return** $k < \max(50 - i, 38 + \delta)$
5   **else if** $o = 1$ **then**
6 | $h \leftarrow 22 + \delta$ // hold with a $\geq$ 22 lead with 4 dice
7 | **if** $i \geq 10 \vee j \geq 23$ **then**
8 | | // if player 1 / 2 has scored 10 / 23, resp.
9 | | $h \leftarrow \max(50 - i, h)$ // then at least roll for the goal
10 | **end if**
11 | **return** $k < h$
12   **else if** $i + j \geq 71$ **then**
13 | // reach the goal when the player score sum reaches 71
14 | **return** $k < 50 - i$
15   **else**   // hold at 5 or goal with 3 dice
16 | **return** $k < \min(50 - i, 5)$
17   **end if**
18 **else**   // player 2 cases
19   **if** $j \geq 50$ **then**   // player 2 must exceed player 1
20 | **return** $k \leq \delta$
21   **else if** $o = 0$ **then**   // keep rolling with 5 dice
22 | **return** $true$
23   **else if** $o = 1$ **then**   // with 4 dice
24 | **if** $i \geq 20 \vee j \geq 32$ **then**
25 | | // if player 1 / 2 has scored 20 / 32, resp.
26 | | **return** $k < 50 - i$ // then roll for the goal
27 | **else**   // else hold with $\geq$ 28 lead
28 | | **return** $k < 18 + \delta$
29 | **end if**
30   **else if** $i + j \geq 84$ **then**
31 | // reach the goal when the player score sum reaches 84
32 | **return** $k < 50 - i$
33   **else**   // hold at 5 or goal with 3 dice
34 | **return** $k < \min(50 - i, 5)$
35   **end if**
36 **end if**

---

8   Conclusions

In this paper, we have computed and visualized optimal play for the 2-player case of the Great Rolled Ones jeopardy dice game. We determined that the first player should start with 3 points for fairest play. In visualizing roll-hold boundaries, we showed a number of interesting nonlinear features, and gave visual insight to the play implications of player 2's advantage from always having the last turn.

In addition, we presented a variety of human-playable strategies, ranging from trivial to complex, with optimal play performance gaps ranging from ~5.4% to ~1.0%, respectively. For casual play, we are especially pleased to recommend Algorithm 3 with an optimal play performance gap of only ~2.0%.

The Great Rolled Ones game has a fairly complex optimal roll-hold policy boundary, as shown in Fig. 1, and yet relatively simple human-playable policies offer decent performance against optimal play, revealing some of the key considerations for excellent play.

References

1. Zombie dice. https://boardgamegeek.com/boardgame/62871/zombie-dice Accessed 24 May 2023
2. Busche, M., Neller, T.W.: Optimal play of the farkle dice game. In: Winands, M.H.M., van den Herik, H.J., Kosters, W.A. (eds.) ACG 2017. LNCS, vol. 10664, pp. 63–72. Springer, Cham (2017). https://doi.org/10.1007/978-3-319-71649-7_6
3. Knizia, R.: Dice Games Properly Explained. Elliot Right-Way Books, Brighton Road, Lower Kingswood, Tadworth, Surrey, KT20 6TD U.K. (1999)
4. Mitschke, S., Scheunemann, R.: Random Fun Generator. Steve Jackson Games, Inc. (2000)
5. Neller, T.W., Presser, C.G.: Optimal play of the dice game pig. UMAP J. **25**(1), 25–47 (2004)
6. Neller, T.W., Presser, C.G.: Practical play of the dice game pig. UMAP J. **31**(1), 5–19 (2010)

Board Games and Card Games

MCTS with Dynamic Depth Minimax

James Ji[✉] and Michael Thielscher

University of New South Wales, Kensington, Australia
jamesx.ji@gmail.com, mit@unsw.edu.au

Abstract. Hybrid models combining Monte-Carlo Tree Search (MCTS) with fixed depth minimax searches have shown great success as the brute force search allow the model to navigate highly tactical domains. However, minimax is computationally expensive and unnecessary in positions that do not require precise calculations. Ideally, we can adjust the depth to efficiently rely on minimax only when needed. In this paper, we build up the motivation for augmenting MCTS with *dynamic* depth minimax searches. We analyse the nature of different domains to create some simple dynamic depth adjustment functions which we then benchmark to reinforce our hypothesis that dynamic adjustments of the search depth in MCTS-Minimax hybrids result in stronger play. For this paper we assume that heuristics or evaluator functions are not available to the player, e.g. as in the context of General Game Playing.

1 Introduction

MCTS describes an application of random-sampling in guiding the growth of a decision tree [1]. It evaluates positions by back propagating the results of simulations. Its evaluations become more accurate over time, eventually converging to optimal play [2]. It came to special prominence in 2016 after Google Deepmind's AlphaGo used a combination of MCTS and neural networks [3] to defeat the 9-dan player Lee Sedol in a five-game match [4]. The inherent randomness of MCTS makes it weak in tactical domains where it takes time to discover narrow, decisive lines. A similar concept is referred to as *traps* [5], where a *level-k* search trap is when the opponent has a winning refutation at most k plies deep. MCTS has also been characterised as making *optimistic moves* [6], as some moves initially seem strong due to a wide range of winning lines but can be refuted with precise play.

Baier and Winands [7,8] discovered an effective approach to this dilemma by embedding shallow minimax searches throughout *MCTS-Minimax Hybrids*. Their models could immediately minimax the surrounding search space to detect decisive lines of play. However, their original paper experimented only with fixed-depth minimax searches. This meant that costly minimax searches were run regardless of the game state, even if no decisive lines would be detected within the search depth. If there are no decisive lines, computation is wasted.

This paper explores the motivation behind augmenting MCTS with dynamically depth-adjusted minimax searches to efficiently take advantage of the tactical strengths of minimax only when the position requires so. We show how the

© The Author(s), under exclusive license to Springer Nature Switzerland AG 2024
M. Hartisch et al. (Eds.): ACG 2023, LNCS 14528, pp. 63–75, 2024.
https://doi.org/10.1007/978-3-031-54968-7_6

frequency of decisive lines can be analysed and suggest some simple functions that adjust the minimax depth accordingly. We then benchmark our dynamic models against the strongest fixed depth models on the same domains as the original paper. Despite the simplicity of our adjustment methods and the strength of our opponent, we achieve strong results in many settings, demonstrating the promise of the dynamic approach alongside the potential to further improve with less crude adjustment methods.

2   Background

MCTS-Minimax Hybrids as first defined by Baier and Winands [7] describes hybrid models which combine standard MCTS with UCT and minimax to significantly outperform MCTS-Solver, a UCT variant that backpropagates proven results [9]; we will revisit their results in Sect. 4.1. In this paper, we will focus on minimax embedded at the selection and expansion phase (MCTS-MS). Before outlining the prior research in this area, we will first motivate the need for such hybrid models. We assume the reader is familiar with MCTS and Minimax.

Tactical Weakness of MCTS. Below displays a position from Chess with White to move.

White is at a piece disadvantage but can play Re8# (Rook to e8, checkmate) to win. However, since MCTS randomly selects children, it might initially select Re2 (Rook to e2), say, giving Black time to defend. To add to the confusion, the randomness of MCTS might lead Black to reply with Nc3 (Knight to c3). If White now follows with Re8#, MCTS will backpropagate this as a win, resulting in an incorrect evaluation of the child node, Re2.

MCTS will eventually evaluate Re8# as the clear winning move; however, the time frame required might be beyond reasonable for the game settings.

Minimax in the Selection and Expansion Phases. When a node satisfying some criteria is encountered during the selection and expansion stages, a fixed-depth minimax search is run and if possible, decisively evaluates the state. Baier and Winands used a visit count threshold as the criterion, but they note that any criterion can be used. This theoretically improves MCTS by guiding the tree growth to avoid shallow losses and detect shallow wins. Such hybrids would easily detect decisive lines such as Re8# in the Chess example. They refer to this as MCTS-MS-d-Visit-v, where d refers to the depth of the minimax and v refers to the visit count required for a node to trigger a minimax search.

The below figure, taken from [7], illustrates the MCTS-MS hybrid. (a) Promising nodes are selected. (b) A node satisfying the minimax criterion is selected. (c) Minimax is run to detect proven game results. (d) If the node's value can be proven, the result is backpropagated. Otherwise, selection continues as normal.

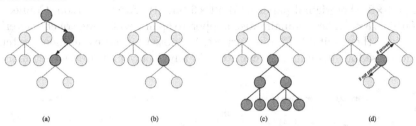

In this paper, we use MCTS-MS-d-Visit-2 as the baseline hybrid model for benchmarking. The original paper used an augmented minimax enhanced with $\alpha\beta$ pruning [10], and in our paper we will also use the enhanced minimax.

3   Dynamic Depth Minimax

Without an external evaluator function, as is assumed in this paper, minimax can only back propagate useful information if it reaches terminal states. In circumstances where terminal states lie outside the search depth, computation resources are simply wasted, leading to poorer performance. We hypothesise that models combining MCTS and minimax can be improved by intelligently allocating computation to minimax, relying on it more in tactical scenarios as in the Chess example in Sect. 2, where it is likely to discover terminal states. We will refer to moves that lead to a solvable position within a given maximum depth as a *terminal line*. Intuitively, we should increase the minimax depth for positions containing many terminal lines and decrease it otherwise.

3.1   Analysis of Terminal Lines

The problem lies in determining whether or not we are in a region with many terminal lines. There are many ways to approach this; here, we analyse the frequency of terminal lines in correlation to three independent variables: turn number, branching factor and rollout length.

Test Conditions. For a given domain, MCTS-Solver played itself for 400 games with 1 s per move. Before starting the timer for each turn, we recorded the number of terminal lines and the aforementioned independent variables.

Measurements. The terminal lines were measured by counting the number of children which could be solved by minimax within a specified maximum depth.

Terminal Ratio. For each turn, we summed the total number of terminal lines across all ongoing games and divided it by the number of ongoing games to produce the *terminal ratio*. Note that some games will end early while others have not reached a conclusion at a given turn number; we refer to the latter as "ongoing games." The current turn number was recorded. Branching factor was recorded as the number of children of the root node. Rollout length was recorded as the average length of 1000 simulations from the root node.

We experimented on the four domains used in the original paper [7]: Breakthrough 8×8 (the original paper used a 6×6 board), Catch the Lion, Connect-4 and Othello 8×8. These domains also happen to span a wide range of characteristics making them suitable for our purpose. We note that our experiments extend beyond these four domains and can work in any 2-player turn-based setting.

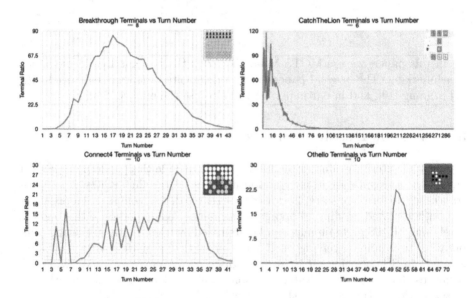

Fig. 1. Terminal Ratio vs Turn Number for the four games

3.2 Analysis and Discussion

Figures 1–3 show the Terminal Ratio measured against turn number, average branching factor and average rollout length. Note that the x-axis does not always span the entire possible domain because not all values appeared in the test. We evaluate the Terminal Ratio as 0 for such cases.

In Fig. 1, the terminal ratio of Connect4 and Breakthrough increases throughout the game because every turn progresses the game towards the end where terminal states are found. They drop off later on where we might expect them to increase due to more terminal states being encountered. However, this is

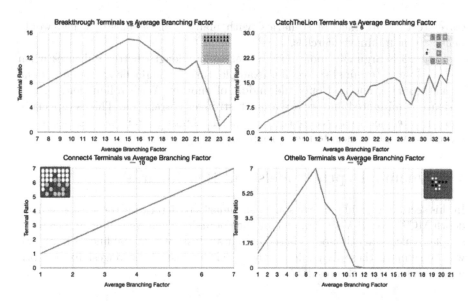

Fig. 2. Terminal Ratio vs Average Branching Factor for the four games

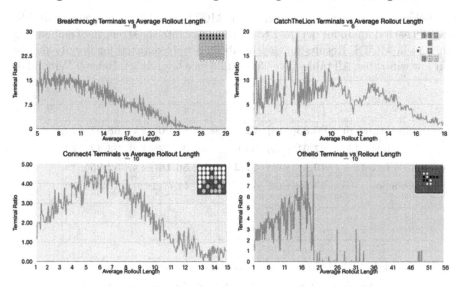

Fig. 3. Terminal Ratio vs Average Rollout Length for the four games

explained by the terminal ratio being an absolute measure, decreasing because there are fewer moves to make.

Figure 2 shows a very strong correlation between the branching factor and Connect4 and Catch the Lion. In these games, more possible lines of play might correlate to more tactical scenarios, with many decisive moves, which would benefit from minimax searches.

Figure 3 is arguably more accurate as rollout length better indicates when we are close to terminal states. For example, turn 30 might be at the end of one Breakthrough game and only the middle of another. With rollout length, this variance is mitigated and we see a much stronger correlation. This is especially true in Catch the Lion, where turn number is almost meaningless as pieces can move backwards to negate their progress.

4 Dynamic Depth Models and Benchmarks

In this section, we will perform some benchmarks to determine the strongest fixed depth MCTS-MS-d-Visit-2 model for each domain. Then we will introduce some simple dynamic depth models based on our analysis of terminal lines. Finally, we will benchmark these dynamic depth models against the strongest fixed depth model and analyse the results.[1]

4.1 Strongest Fixed Depth Model

The below tables show the win-rates of MCTS-MS-d-Visit-2 against MCTS-Solver in each domain for depths 1 to 8 to demonstrate the strength of embedding minimax in MCTS. Results for depths above 8 were omitted for brevity due to their low win-rates. All tables give 95% confidence bounds (cf. Baier & Winands).

	1	2	3	4
Breakthrough	**60.83**(±4.8)	**72.50**(±4.4)	**71.00**(±4.4)	**68.67**(±4.5)
Catch the Lion	**87.91**(±3.2)	**93.69**(±2.4)	**94.22**(±2.3)	**93.42**(±2.4)
Connect-4	**49.45**(±4.9)	**51.28**(±4.9)	**56.16**(±4.9)	**52.95**(±4.9)
Othello	**55.59**(±4.9)	43.01(±4.9)	44.77(±4.9)	37.99(±4.8)

	5	6	7	8
Breakthrough	28.00(±4.4)	14.00(±3.4)	18.50(±3.8)	13.00(±3.3)
Catch the Lion	**92.95**(±2.5)	**87.59**(±3.2)	**58.63**(±4.8)	31.36(±4.5)
Connect-4	38.97(±4.8)	31.41(±4.5)	15.35(±3.5)	17.28(±3.7)
Othello	31.55(±4.6)	20.21(±3.9)	19.90(±3.9)	9.28(±2.8)

---

[1] We used a 2013 iMac, 2.7 Ghz Quad-Core Intel Core i5, 16 GB RAM, 1600 MHz DDR3. We played 200 games per side for each match with 1 s per move. We played more games to differentiate close results as needed. Drawn games (which are infrequent) are discarded. Win rates are given with 95% confidence intervals.

We then determined the strongest depth for MCTS-MS-d-Visit-2 in each domain via round robin. The below tables show the win-rates of the strongest model for each domain against all other depths 1 to 8 to provide an idea of their relative strength. The strongest fixed depth model in each domain is as follows: Breakthrough: 3, Catch the Lion: 4, Connect-4: 3, Othello: 1.

	1	2	3	4
Breakthrough	**67.25**($\pm$4.6)	**63.50**($\pm$4.7)	–	**50.75**($\pm$4.9)
Catch the Lion	**80.05**($\pm$3.9)	**71.21**($\pm$4.4)	**57.14**($\pm$4.8)	–
Connect-4	**54.96**($\pm$4.9)	48.21($\pm$4.9)	–	**52.67**($\pm$4.9)
Othello	–	**53.28**($\pm$4.9)	**50.65**($\pm$4.9)	**59.63**($\pm$4.8)

	5	6	7	8
Breakthrough	**77.50**($\pm$4.1)	**90.00**($\pm$2.9)	**88.50**($\pm$3.1)	**91.50**($\pm$2.7)
Catch the Lion	**68.01**($\pm$4.6)	**73.74**($\pm$4.3)	**85.00**($\pm$3.5)	**89.22**($\pm$3.0)
Connect-4	**65.31**($\pm$4.7)	**73.71**($\pm$4.3)	**85.57**($\pm$3.4)	**89.19**($\pm$3.0)
Othello	**65.42**($\pm$4.7)	**73.02**($\pm$4.4)	**83.55**($\pm$3.6)	**89.97**($\pm$2.9)

Note that occasionally the strongest model might perform slightly worse against a model of another depth. However, they performed the best overall in round robin. This satisfies our purpose of picking a strong fixed depth model to benchmark against.

4.2   Adjusting Dynamic Depth Models

Intuitively, we should adjust our depth according to our analysis, increasing and decreasing the depth when the terminal ratio is high and low. A simple approach involves directly correlating the minimax depth with the terminal ratio throughout the game. The minimax depth then needs to be linearly scaled to an appropriate range to ensure the experiment can run within the time constraints.

We created four simple depth adjustment functions which closely follow the graphs from our analysis. We will use LS to represent a function linearly normalised to the range $[0, 1]$. Let the graphs in Fig. 1, Fig. 2 and Fig. 3 be denoted as functions, f_{turn}, f_{branch} and f_{roll} respectively.

Turn-A $LS(f_{turn})$ Simple linearly normalised function on turn number.

Turn-B $LS(LS(f_{turn}) \cdot LS(t))$ where t denotes the current turn number. This essentially puts more weighting on later turn numbers, thus increasing minimax depth more during later turns.

Branch-A $LS(f_{branch})$. Simple linearly normalised function on average branching factor.

Roll-A $LS(f_{roll})$. Simple linearly normalised function on average rollout length.

For each depth adjustment function we will create a dynamic depth model by replacing the d in MCTS-MS-d-Visit-2 with our own depth adjustment function. We also benchmarked multiple weightings for each depth adjustment function like so: $f_{dynamic} \cdot k$ for $k \in \mathbf{Z}, 1 \leq k \leq d_{max}$ where $f_{dynamic}$ denotes any depth adjustment function and d_{max} denotes the maximum minimax depth used for that domain. Since we tested multiple depths to determine the strongest fixed depth model, it is fair to test multiple weights for our depth adjustment function. For convenience, we named our models following the convention **FUNCTION-k**. For example, the model equipped with depth adjustment function **TURN-A** with weighting $k = 5$ is referred to as **TURN-A-5**.

4.3 Analysis of Results

Turn-A. In Connect-4 the dynamic depth model outperforms when $k = 4$. This suggests that by varying the depth of the minimax searches to be mostly shallow but occasionally going deeper than the fixed depth model ($d = 3$), we can achieve stronger results, reinforcing our hypothesis of efficient minimax usage. Our model also shows strong performance in Othello for many k. This matches our intuition that fixed depth minimaxes would be useless early in Othello as the game only nears completion when the board is filled. There are some promising results in Breakthrough ($k = 2, 4$). Performance in Catch the Lion is poor across the board, however. This could be due to the small board space of Catch the Lion, requiring precise minimaxes early on to decisively maneuver into a winning position. The tables below shows the win-rate of TURN-A-k vs the strongest fixed depth model in each domain for $1 \leq k \leq d_{max}$.

	1	2	3	4	5
Breakthrough (3)	37.25(±4.7)	**49.25**(±4.9)	48.25(±4.9)	**49.25**(±4.9)	44.5(±4.9)
Catch the Lion (4)	40.83(±4.8)	43.18(±4.9)	42.82(±4.8)	39.67(±4.8)	42.79(±4.8)
Connect-4 (3)	46.24(±4.9)	41.37(±4.8)	42.66(±4.8)	**51.37**(±4.9)	**49.87**(±4.9)
Othello (1)	**52.00**(±4.9)	**49.13**(±4.9)	48.38(±4.9)	**49.50**(±4.9)	48.13(±4.9)

	6	7	8	9	10
Breakthrough (3)	27.5(±4.4)	26.5(±4.3)	19.25(±3.9)	–	–
Catch the Lion (4)	33.45(±4.6)	–	–	–	–
Connect-4 (3)	45.08(±4.9)	46.87(±4.9)	42.51(±4.8)	46.61(±4.9)	32.47(±4.6)
Othello (1)	46.63(±4.9)	**53.25**(±4.9)	**49.88**(±4.9)	**51.63**(±4.9)	45.75(±4.9)

Turn-B. By weighing the depth more so that later turns trigger deeper mini-maxes, we achieve even stronger results in Othello, notably at $k = 10$. This is due to turn number being a very strong signal in this domain since terminal lines are only found in the end game. This reinforces our hypothesis that intelligent usage of minimax is important. This also achieves comparable performances to the fixed depth model in Breakthrough $k = 3$ and Connect-4 $k = 8, 9$. The tables below show the win-rate of TURN-B-k vs the strongest fixed depth model in each domain for $1 \leq k \leq d_{max}.1 \leq k \leq d_{max}$.

	1	2	3	4	5
Breakthrough (3)	38.50(±4.8)	45.83(±4.9)	**49.67**(±4.9)	48.92(±4.9)	40.00(±4.8)
Catch the Lion (4)	11.50(±3.1)	31.82(±4.6)	23.06(±4.1)	18.50(±3.8)	13.57(±3.4)
Connect-4 (3)	46.70(±4.9)	46.13(±4.9)	46.17(±4.9)	46.22(±4.9)	40.32(±4.8)
Othello (1)	**52.05**(±4.9)	**50.81**(±4.9)	**51.24**(±4.9)	**52.12**(±4.9)	48.32(±4.9)

	6	7	8	9	10
Breakthrough (3)	26.75(±4.3)	18.50(±3.8)	15.5(±3.5)	–	–
Catch the Lion (4)	11.50(±3.1)	–	–	–	–
Connect-4 (3)	44.47(±4.9)	48.33(±4.9)	**49.07**(±4.9)	**50.68**(±4.9)	42.01(±4.8)
Othello (1)	**49.18**(±4.9)	**49.87**(±4.9)	**53.39**(±4.9)	**51.09**(±4.9)	**54.57**(±4.9)

Branch-A. Connect-4 achieves an incredibly strong result at $k = 3$. The reason could be due to the linear relationship between branching factor and terminal ratio shown in Fig. 2, suggesting the former is a strong predictor of the lat-ter. This suggests that the fixed-depth model $d = 3$ was wasting unnecessary minimax searches at the cost of performance. Results in Othello remain strong across the board for reasons outlined earlier. The tables below show the win-rate of BRANCH-A-k vs the strongest fixed depth model in each domain for $1 \leq k \leq d_{max}.1 \leq k \leq d_{max}$.

	1	2	3	4	5
Breakthrough (3)	40.00(±4.8)	41.50(±4.8)	47.00(±4.9)	48.25(±4.9)	24.50(±4.2)
Catch the Lion (4)	8.88(±3.0)	10.25(±3.0)	18.14(±3.7)	21.75(±4.0)	27.85(±4.4)
Connect-4 (3)	45.00(±4.9)	39.11(±4.8)	**59.56**(±4.8)	**50.79**(±4.9)	45.16(±4.9)
Othello (1)	**54.40**(±4.9)	**50.55**(±4.9)	**50.41**(±4.9)	48.53(±4.9)	48.91(±4.9)

	6	7	8	9	10
Breakthrough (3)	8.25(±3.7)	10.25(±3.0)	10.50(±3.0)	–	–
Catch the Lion (4)	42.96(±4.9)	–	–	–	–
Connect-4 (3)	34.57(±4.7)	31.15(±4.5)	13.37(±3.3)	12.37(±3.2)	5.00(±2.1)
Othello (1)	**51.11**(±4.9)	48.61(±4.9)	**53.51**(±4.89)	**53.33**(±4.9)	47.83(±4.9)

Roll-A. The results are disappointing for all domains apart from Othello. This could be explained by the high variance in our average rollout length measurements. Since our model minimaxed on the second visit, our average rollout length was sampled from just two random rollouts. For fairness, sampling more rollouts to obtain a more reliable average was not an option as it would increase computation costs, making it impossible to isolate the impact of rollout lengths on the dynamic depth model when benchmarking against the fixed depth model. Alternatively, we could increase the visit threshold (e.g. from 2 to 10) for all models to even the playing field. We leave this to future experimentation. The tables below show the win-rate of ROLL-A-k vs the strongest fixed depth model in each domain for $1 \leq k \leq d_{max}.1 \leq k \leq d_{max}$.

	1	2	3	4	5
Breakthrough (3)	41.25(±4.8)	40.50(±4.8)	42.00(±4.8)	35.25(±4.7)	26.00(±4.3)
Catch the Lion (4)	13.00(±3.3)	17.34(±3.7)	18.80(±3.8)	28.82(±4.4)	29.29(±4.5)
Connect-4 (3)	43.29(±4.9)	44.96(±4.9)	46.92(±4.9)	34.54(±4.6)	39.10(±4.8)
Othello (1)	**52.96**(±4.9)	48.64(±4.9)	**49.31**(±4.9)	**51.60**(±4.9)	**50.00**(±4.9)

	6	7	8	9	10
Breakthrough (3)	7.75(±2.6)	3.50(±1.8)	4.75(±2.1)	–	–
Catch the Lion (4)	20.40(±3.9)	–	–	–	–
Connect-4 (3)	31.56(±4.6)	26.87(±4.3)	12.76(±3.3)	11.96(±3.2)	8.51(±2.7)
Othello (1)	**59.40**(±4.8)	41.42(±4.8)	35.22(±4.7)	34.51(±4.7)	**51.08**(±4.9)

Strong Performance in Othello. Our models achieve strong results in Othello across the board. Othello is the perfect example of a domain where minimax searches are wasted early game and as a result, is almost guaranteed to benefit from dynamic depth adjustments, even with our crude adjustment functions.

Weak Performance in Catch the Lion. Our models achieve the worst results in Catch the Lion. A likely explanation is that our strongest fixed depth model in Catch the Lion is already very strong as shown in the tables at the end of Sect. 4.1, where its win-rate against other fixed depth models range from a

TURN-A	1	2	3	4	5
Breakthrough (3)	**61.00**(±4.8)	**62.75**(±4.7)	**67.5**(±4.6)	**69.00**(±4.5)	**50.50**(±4.9)
Catch the Lion (4)	**84.89**(±3.5)	**86.87**(±3.3)	**92.17**(±2.6)	**92.17**(±2.6)	**90.93**(±2.8)
Connect-4 (3)	**52.87**(±4.9)	45.19(±4.9)	**52.81**(±4.9)	49.46(±4.9)	**56.25**(±4.9)
Othello (1)	**49.59**(±4.9)	**49.59**(±4.9)	47.43(±4.9)	**49.46**(±4.9)	**49.46**(±4.9)
TURN-A	6	7	8	9	10
Breakthrough (3)	**51.75**(±4.9)	**39.25**(±4.8)	38.75(±4.8)	–	–
Catch the Lion (4)	**92.19**(±2.6)	–	–	–	–
Connect-4 (3)	46.51(±4.9)	**53.68**(±4.9)	46.74(±4.9)	**52.69**(±4.9)	48.91(±4.9)
Othello (1)	45.71(±4.9)	**49.28**(±4.9)	**53.62**(±4.9)	47.83(±4.9)	46.43(±4.9)

TURN-B	1	2	3	4	5
Breakthrough (3)	**61.00**(±4.8)	**62.75**(±4.7)	**67.50**(±4.6)	**69.00**(±4.5)	**50.50**(±4.9)
Catch the Lion (4)	**84.00**(±3.6)	**93.39**(±2.4)	**84.00**(±3.6)	**95.96**(±1.9)	**95.88**(±1.9)
Connect-4 (3)	**53.41**(±4.9)	47.73(±4.9)	**51.14**(±4.9)	**60.00**(±4.8)	**56.82**(±4.9)
Othello (1)	**54.57**(±4.9)	**51.09**(±4.9)	**53.59**(±4.9)	49.87(±4.9)	49.18(±4.9)
TURN-B	6	7	8	9	10
Breakthrough (3)	**51.75**(±4.9)	39.25(±4.8)	34.25(±4.7)	–	–
Catch the Lion (4)	**93.00**(±2.5)	–	–	–	–
Connect-4 (3)	**54.65**(±4.9)	**52.81**(±4.9)	**52.75**(±4.9)	**55.56**(±4.9)	**53.41**(±4.9)
Othello (1)	48.38(±4.9)	**52.11**(±4.9)	**51.24**(±4.9)	**50.81**(±4.9)	**52.05**(±4.9)

BRANCH-A	1	2	3	4	5
Breakthrough (3)	**58.67**(±4.8)	**68.67**(±4.5)	**75.33**(±4.2)	**68.67**(±4.5)	**50.67**(±4.9)
Catch the Lion (4)	**84.5**(±3.5)	**83.76**(±3.6)	**87.94**(±3.2)	**90.95**(±2.8)	**93.91**(±2.3)
Connect-4 (3)	**53.41**(±4.9)	47.73(±4.9)	**51.14**(±4.9)	**60.00**(±4.8)	**56.82**(±4.9)
Othello (1)	**50.38**(±4.9)	**51.43**(±4.9)	48.18(±4.9)	43.51(±4.9)	42.86(±4.8)
BRANCH-A	6	7	8	9	10
Breakthrough (3)	20.00(±3.9)	14.67(±3.5)	13.33(±3.3)	–	–
Catch the Lion (4)	**92.96**(±2.5)	–	–	–	–
Connect-4 (3)	**54.65**(±4.9)	**52.81**(±4.9)	**52.75**(±4.9)	**55.56**(±4.9)	**53.41**(±4.9)
Othello (1)	39.01(±4.8)	35.21(±4.7)	21.92(±4.05)	19.73(±3.9)	16.08(±3.6)

ROLL-A	1	2	3	4	5
Breakthrough (3)	10.25(±3.0)	7.82(±4.0)	17.25(±3.7)	43.25(±4.9)	**61.75**(±4.8)
Catch the Lion (4)	**86.50**(±3.3)	**91.90**(±2.7)	**92.73**(±2.5)	**92.15**(±2.6)	**90.48**(±2.9)
Connect-4 (3)	9.95(±2.9)	16.36(±3.6)	31.49(±4.5)	34.29(±4.7)	38.11(±4.8)
Othello (1)	**49.35**(±4.9)	44.12(±4.9)	38.92(±4.8)	44.05(±4.9)	48.36(±4.9)
ROLL-A	6	7	8	9	10
Breakthrough (3)	**68.00**(±4.6)	**57.00**(±4.9)	**65.25**(±4.7)	–	–
Catch the Lion (4)	**88.13**(±3.2)	–	–	–	–
Connect-4 (3)	44.17(±4.9)	**52.66**(±4.9)	**55.83**(±4.9)	**52.04**(±4.9)	**52.96**(±4.9)
Othello (1)	**50.42**(±4.9)	46.59(±4.9)	**49.03**(±4.9)	**50.28**(±4.9)	**49.02**(±4.9)

Fig. 4. Win-rate of TURN-A, TURN-B, BRANCH-A, ROLL-A vs MCTS-Solver in each domain for $1 \leq k \leq d_{max}$

minimum (but still very high) 57.14% to as high as 80.05%. More precise dynamic depth functions might be needed to outperform the strongest fixed depth model.

Results vs MCTS-Solver. The purpose of Fig. 4 is to show that our model is strong generally and not only against the fixed depth model. Indeed we see strong performances across the board from our dynamic depth models. Note that whilst the results for Othello might seem unimpressive, the fixed depth model also sees the same issue. As mentioned previously in this paper, this domain is one in which minimax searches must be used very carefully as they are strictly useless for most of the game.

5   Conclusion

Dynamic depth minimax models are promising, occasionally matching or surpassing the strongest fixed depth model. Their strong performance in domains such as Othello show that intelligent minimax usage is important. However, the key takeaway is while the strength of the fixed depth models are capped as there is no room for change, the dynamic depth models have potential for improvement. Note that we used extremely rudimentary adjusment functions and we also benchmarked against the strongest fixed depth models. It is not a stretch to assume that with better adjustment functions, we can achieve even stronger performances.

We also leave to future research other novel approaches for analysing the nature of terminal lines which could lead to new ideas on dynamic depth models. In this paper, we used turn number, branching factor and rollout length. However, other heuristics such as number of pieces on the board, or even implicit heuristics are conceivable.

References

1. Coulom, R.: Efficient selectivity and backup operators in Monte-Carlo tree search. In: 5th International Conference on Computer and Games, pp. 72–83 (2006)
2. Koscis, L., Szpesvári, C.: Bandit based Monte-Carlo planning. In: 17th European Conference on Machine Learning (ECML), pp. 282–293 (2006)
3. Silver, D., Huang, A., Maddison, C., et al.: Mastering the game of Go with deep neural networks and tree search. Nature **529**, 484–489 (2016)
4. The Guardian, https://www.theguardian.com/technology/2016/mar/15/googles-alphago-seals-4-1-victory-over-grandmaster-lee-sedol. Accessed 9 Nov 2023
5. Ramanujan, R., Sabharwal, A., Selman, B.: On adversarial search spaces and sampling-based planning. In: 20th International Conference on Automated Planning and Scheduling (ICAPS), pp. 242–245 (2010)
6. Finnson, H., Björnsson, Y.: Game-tree properties and MCTS performance. In: IJCAI Workshop on General Intelligence in Game Playing Agents, pp. 23–30 (2011)
7. Baier, H., Winands, M.: Monte-Carlo tree search and minimax hybrids. In: IEEE Conference on Computational Intelligence in Games (CIG), pp. 1–8 (2013)

8. Baier, H., Winands, M.: MCTS-Minimax hybrids. IEEE Trans. Comput. Intell. AI Games **7**(2), 167–179 (2015)

9. Winands, M., Björnsson, Y., and Saito, J.: Monte-Carlo tree search solver. In: 6th International Conference on Computers and Games, pp. 25–36 (2008)

10. Knuth, D.E., Moore, R.W.: An analysis of alpha-beta pruning. Artif. Intell. **6**(4), 293–326 (1975)

Can We Infer Move Sequences in Go from Stone Arrangements?

Chu-Hsuan Hsueh[(✉)] [ID] and Kokolo Ikeda

Japan Advanced Institute of Science and Technology, Nomi, Ishikawa, Japan
{hsuehch,kokolo}@jaist.ac.jp

Abstract. Inference commonly happens in our daily lives and is also a hot topic for AI research. In this paper, we infer move sequences in Go, i.e., the order in which moves are played, from stone arrangements on the board. We formulate the problem as likelihood maximization and employ a general optimization algorithm, simulated annealing, to solve it. Our experiments on professional and amateur games show that the proposed approach sometimes produces more natural move sequences than those played by humans.

Keywords: Inference · likelihood maximization · simulated annealing · Go

1 Introduction

Inference is a kind of intelligent work widely done by humans, meaning guesses or opinions based on known information. For example, detectives try to infer criminals and the approaches from pieces of evidence left at the scenes of the crimes. Doctors try to infer patients' abnormalities and the causes from indirect information such as symptoms and body checks. Inference also takes place in games. A typical example is to infer the cards on the opponents' hands in poker.

For the game of Go, strong human players can infer the move sequences at a glance of board states. There are two main reasons. First, most stones are on the board unless captured. Second and more importantly, strong players have played many games and know how games usually proceed. It is interesting to investigate whether and how computers can tackle such inference tasks.

In this paper, we formulate as a likelihood maximization problem the inference of a Go game's move sequence from stones on the board. An example on 9×9 boards is shown in Fig. 1, while we actually work on 19×19 boards. Assuming that the likelihood function of human players' moves is available, we treat the problem of finding the most likely move sequence as finding the move sequence with the highest likelihood. To simplify the problem, we target opening games without capturing stones in this paper.

This work was supported by JSPS KAKENHI Grant Numbers JP23K17021 and JP23K11381.

© The Author(s), under exclusive license to Springer Nature Switzerland AG 2024
M. Hartisch et al. (Eds.): ACG 2023, LNCS 14528, pp. 76–87, 2024.
https://doi.org/10.1007/978-3-031-54968-7_7

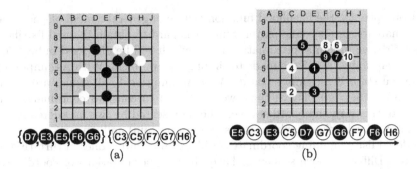

Fig. 1. Examples of (a) a stone arrangement and (b) a move sequence.

We employed simulated annealing (SA) [4], a general optimization algorithm often used for discrete search spaces, to solve our problem. We experimented with both professional and amateur players' games. We used a neural network trained using strong human players' games to approximate the likelihood of move sequences. The results showed that our approach inferred professional games better than amateur games and inferred shorter move sequences better than longer ones. Interestingly, for both professional and amateur games, our approach sometimes produced move sequences more natural than human players'.

The rest of this paper is structured as follows. Section 2 explains the problem formulation. Section 3 describes the background, including related work and SA. Section 4 presents our approaches for move sequence inference, and the results are shown in Sect. 5. Finally, Sect. 6 makes concluding remarks.

2 Problem Formulation

We formulate the move sequence inference in Go as a likelihood maximization problem[1]. The input is the stone arrangement on the board (e.g., Fig. 1a), and the output is the move sequence (e.g., Fig. 1b). First, we make three assumptions:

1. The likelihood function of human players' moves is available.
2. There are no handicap stones (i.e., black stones on the initial board).
3. No stones are captured.

The likelihood function in Assumption 1 is represented by a parameter θ, which outputs the probability that human players play a move m at a state s, denoted by $P_\theta(m|s)$, for all state-move pairs. For a move sequence (or history) $h = (m_1, m_2, ..., m_n)$ with n moves, the likelihood is calculated by $L_\theta(h) = \Pi_{i=1}^n P_\theta(m_i|s_{i-1})$, where s_0 is the initial state and s_i the state after playing m_1 to

---

[1] In some cases, players may intentionally play unexpected moves (e.g., to transpose to openings that are not well-studied so that the opponent may make mistakes). Such cases are not the inference targets in this paper.

m_i. Conceptually, the likelihood function can be of general human players, mid-level amateurs, or a specific player. Since it is hard to obtain the real likelihood in practice, we use an approximation for θ, which will be discussed in Sect. 5.

Assumptions 2 and 3 are made to simplify the problem. With Assumption 2, the initial state is an empty board. With Assumption 3, all stones are on the board after they are played; thus, we only need to consider coordinates with stones. In this way, the board state can be represented by two unordered sets of coordinates for black and white stones (Fig. 1a bottom) and a move sequence by an ordered list containing coordinates of black and white stones in turn (Fig. 1b bottom). Different move sequences have different permutations of coordinates, constraining that black and white stones are put one by another.

The problem of finding the most likely move sequence is then regarded as the problem of finding a permutation of stone coordinates with the highest likelihood. With n moves, the number of permutations is $(\lceil n/2 \rceil)!(\lfloor n/2 \rfloor)!$, indicating that this is a nontrivial problem.

3   Background

Subsection 3.1 discusses related work and Subsect. 3.2 introduces SA.

3.1   Related Work

Researchers have applied likelihood maximization to solve various types of inference problems. For example, it has been widely used to reconstruct phylogenies in biology (the evolutionary history of organisms) [2]. Also, it has been applied to infer correct answers of tasks from the results of crowdsourcing platforms [1].

Some researchers built systems to record played moves in Go from videos (e.g., TV programs) [3,8]. They used image-processing techniques to identify boards and stones and then extracted moves when changes occurred. Their systems also generated move sequences from stone arrangements, but the problem and approaches differed from ours.

3.2   Simulated Annealing

Simulated annealing (SA) [4] is a stochastic optimization algorithm often used for problems of discrete search spaces. Starting with an initial solution, SA repeats the following processes for some iterations: (i) creating a new solution s' by a neighbor function that alters the current solution s a little, (ii) evaluating s and s' by a cost function that gives lower values to better solutions, and (iii) replacing s with s' with a probability of 100% when s' is better than s or a certain probability otherwise. The probability is $\exp(-\Delta E/T)$, where ΔE is calculated by the cost of s' minus that of s, and T is a temperature that gradually decreases over iterations. Users need to decide the initial and the final temperatures, the number of iterations, the solution representation (including initialization), and neighbor and cost functions, but nothing else.

4 Simulated Annealing for Inferring Go Move Sequences

The following explains our solution initialization and cost and neighbor functions. A trivial way to obtain an initial solution is to randomly decide the order of stone coordinates in the list. In addition to this, we propose a *greedy* method that uses the likelihood function as a heuristic. In more detail, we start from the initial state, select an unused stone of the player to move with the highest likelihood, play this move, and repeat this process until all stones are used.

Regarding the cost function, given a move sequence, we calculate its geometric mean of likelihood and define its cost as the inverse of the geometric mean so that better solutions have lower values. Calculating the geometric means is to make the cost function general to move sequences with different lengths. Otherwise, the likelihood usually gets smaller as the move sequence gets longer because the probability of playing a move at a state is usually lower than 1. Taking a move sequence with 20 moves and a likelihood of 10^{-40} as an example, the cost is $1/((10^{-40})^{1/20}) = 100$.

Neighbor functions in SA are usually designed to alter a solution a little, attempting to progressively improve the solutions through iterations. We propose 4 neighbor functions: **swp2**–randomly selecting 2 stones of the same color and swapping them (Fig. 2a), **mov2**–randomly selecting 2 consecutive stones and moving them to some other places while constraining that no stones of the same colors are put together (Fig. 2b), **rot3**–randomly selecting 3 stones of the same color and rotating them (Fig. 2c), and **mix**–using the above three alternately.

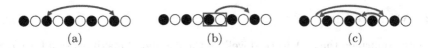

(a)   (b)   (c)

Fig. 2. Illustrations of neighbor functions: (a) swp2, (b) mov2, and (c) rot3.

5 Experiments

This section presents our experiments. Subsections 5.1 to 5.4 show the settings, the main results, selected move sequences, and ablation studies on SA options.

5.1 Experiment Settings

Regarding the likelihood function of human players' moves in Assumption 1, we employed a neural network (abbr. NN) trained using strong human players' games[2]. The NN takes the latest 8 states as inputs and then outputs the current

---

[2] https://sjeng.org/zero/best_v1.txt.zip from the Leela Zero project, https://github.com/leela-zero/leela-zero.

state's policy (the probability distributions over moves) and value. We treated the policy output as the approximation of the likelihood function.

For game records used in the experiments, we chose professionals' and mid-level amateurs' games. Since the NN was trained using strong human players' games, we expected the network to be able to reflect the likelihood function of professional players' moves. For comparison, we intentionally included amateurs' games. It is interesting to investigate whether the NN is suitable for inferring weaker players' moves or whether the inferred move sequences look more reasonable than amateur players'. We selected 100 games for each set from openly available datasets[3]. The selected games did not contain handicap stones nor captured stones until the 50th move to satisfy Assumptions 2 and 3. We clipped the length to 20 and 40 moves from each game to test the scalability, while longer games remain as future work. To sum up, we had 4 sets of games: {professional, amateur} $\times$ {20 moves, 40 moves}, abbreviated by {Pro, Ama} $\times$ {20m, 40m}. Although we knew the actual move sequence in each game, the information was removed when making inferences. Namely, the inputs only contained the stone arrangements right after the 20th or 40th move was played (e.g., Fig. 1a bottom).

To infer the move sequence of a stone arrangement, we ran SA 5 times and took the move sequence with the highest likelihood as the solution. It is common to run SA several times because randomness is involved. We decided the parameters of the iteration number to be 5,000 and 20,000 for 20m and 40m games, respectively, and the initial and final temperatures to be 1.0 and 0.1, respectively, by some preliminary experiments. Initial solutions were obtained by the greedy method, and the mix neighbor function was applied.

5.2 Main Results

We first confirmed that SA worked well on the likelihood maximization for move sequence inference. We then investigated how much the inferred move sequences matched the actual ones. For some mismatched move sequences, we found that the inferred ones looked more reasonable than the actual ones. To analyze this, we quantified the goodness of moves. The following presents the details.

Likelihood Maximization. To confirm whether the likelihood maximization was well done, ideally, we should compare the likelihoods of the inferred move sequences with the true maximum likelihoods. However, with 20 or 40 moves, each stone arrangement has $(10!)^2 \approx 10^{13}$ or $(20!)^2 \approx 10^{36}$ possible move sequences, making it infeasible to obtain the one with the maximum likelihood. Thus, we used the likelihood of the actual move sequences as baselines instead.

Figure 3 shows the scatter plots of the geometric means of likelihoods, where the x-axis represents the actual move sequences and the y-axis the inferred ones.

---

[3] Professional: https://mega.co.nz/#!xFE2kTaK!Oj3_N9NpGmYVGTuka7Nc3T0HT mp3kKcXZR6p1Q7U5YU, amateur (1d): https://github.com/featurecat/go-data set.

A point above $y = x$ means that the inferred move sequence has a higher like-lihood than the actual one. Under Assumption 1, the inferred move sequence is more likely played than the actual one. For the set of Pro20m, all points were on or above $y = x$. For the sets of Pro40m, Ama20m, and Ama40m, the ratios of points on or above $y = x$ were 65%, 93%, and 82%, respectively. We also obtained regression lines of the data points, constraining the intercepts to be 0. A slope higher than 1 indicates that the inferred move sequences generally have a higher likelihood than actual ones. The slopes of the 4 sets were 1.023, 0.988, 1.093, and 1.125. We concluded that the optimization was generally well done, though there was room to improve, especially in the Pro40m set.

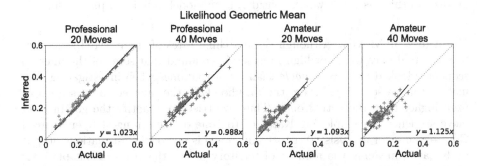

Fig. 3. Scatter plots of the geometric means of likelihoods: Pro20m, Pro40m, Ama20m, and Ama40m from the left.

Matching Between Actual and Inferred Sequences. After confirming that the likelihood maximization was generally well done, we investigated how much the inferred move sequences matched the actual ones. Among the inferred sequences, 51%, 2%, 11%, and 0% exactly matched the actual ones for Pro20m, Pro40m, Ama20m, and Ama40m, respectively. Longer sequences were more chal-lenging to infer, as expected. In addition, it was also as expected that our app-roach matched Pro better than Ama.

With more moves, the matching rates decreased. Two possible reasons are discussed as follows. First, longer games, i.e., more complex solutions, made the optimization of SA more unstable. This was supported by the standard deviations of the likelihood geometric means of the 5 trials, which were 0.020, 0.028, 0.010, and 0.016 for the above 4 sets. Thus, SA might need more trials or tricks (e.g., advanced neighbor functions) to obtain a good solution. Second, when games proceed to the mid-game phase, it might be difficult for players to play good or natural moves. In contrast, in the opening phase, players can follow standard sequences (*joseki*), which is easier to match. Although it was not frequent for our approach to reproduce move sequences that exactly matched the actual ones, we found it successfully reproduced many sub-sequences.

The following discusses the differences between Pro and Ama, i.e., higher likelihoods (Fig. 3) and matching rates for Pro. We approximated the likelihood function of human players' moves by an NN trained using strong human players' games. Thus, we considered it reasonable that the NN could better infer moves in Pro (who belong to strong players) than those in Ama. McIlroy-Young et al. [5] also reported that human players with different skill levels were best predicted by NNs trained using games of the corresponding level in chess. To make better inferences for the Ama cases, it is worth trying to train NNs for the corresponding skill levels. One potential problem we considered was that it might be intrinsically difficult to train an NN to well predict different amateur players of the same skill level because the players play a variety of moves (including bad moves). In this case, it is worth considering to model individual players [6].

Goodness of Moves. Although the matching rates of the inferred sequences to the actual ones were not high in general, we found that some of the inferred sequences looked more *reasonable* than the actual ones when investigating mismatched move sequences where the likelihoods of the inferred move sequences were higher than the actual ones. Thus, we tried to quantify the goodness of move sequences by employing KataGo [10], one of the strongest open-sourced Go programs, for analysis. Territory advantage is one of the estimations output by KataGo, meaning the number of territory points that the current player is leading (e.g., +2.5 means that the current player leads the opponent by 2.5 points in territory). We defined the loss of a move m_i in a move sequence by $\delta_{maxi} - \delta_i$, where δ_i is the territory advantage of m_i and δ_{maxi} the maximum territory advantage among moves at the state before playing m_i. We then defined the loss of a move sequence by the average loss of the moves in the sequence. Move sequences with high losses usually contain some extremely bad moves, which may look unnatural. Thus, we considered move sequences with lower losses to be likely to look more reasonable.

Figure 4 shows the scatter plots of the territory advantage losses, where the x-axis represents the actual move sequences and the y-axis the inferred ones. A point below $y = x$ means that the inferred move sequence has a lower loss than the actual one. Similar to Fig. 3, we obtained the regression lines of the points. In addition, we included the likelihood information in the plots. We calculated the ratio of the likelihood geometric means between the inferred and the actual sequences. A higher ratio means that the likelihood of the inferred sequence is much higher than the actual one. The ratios are represented by colors: higher ratios are closer to red, a ratio of 1 is white, and lower ratios are closer to blue.

The regression lines' slopes for Pro20m, Pro40m, Ama20m, and Ama40m were 1.002, 1.425, 0.903, and 0.929, respectively. A slope higher than 1 indicates that the inferred sequences tend to have higher losses than the actual ones (i.e., the inferred sequences with worse move goodness). Interestingly, the slopes for Pro were > 1 while Ama's were < 1. As shown in Fig. 4, some inferred move sequences in Pro had much higher losses than the actual sequences. In contrast,

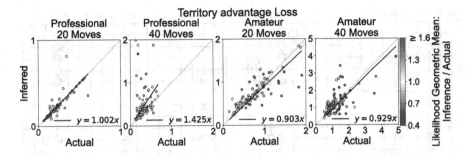

Fig. 4. Scatter plots of the territory advantage loss: Pro20m, Pro40m, Ama20m, and Ama40m from the left, where colors represent the ratio of the likelihood geometric means between the inferred and the actual move sequences. (Color figure online)

the inferred sequences in Ama had relatively close losses to the actual sequences, and some of the losses were much lower.

Moreover, when looking at the colors together, we had two interesting observations, one from the Pro plots and the other from the Ama plots. In the Pro plots, some red or pink points were in the upper-left corners, especially in the 40m case. This means that the likelihood maximization was well done, but some extremely bad moves were produced in the move sequences. Since the goodness of moves was not the target of the maximization problem, it was not strange to have move sequences containing bad moves. As a future research direction, it is worth trying to reformulate the problem to also consider the goodness of moves (e.g., combining the policy outputs from NN trained using human players' games and NN trained using AlphaZero [7]).

In the Ama plots, many deep red points (the likelihood of the inferred sequence being much higher than the actual one) were below $y = x$ (the loss of the inferred sequence being lower than the actual one). We interpreted the results to be that our approach could produce more natural and reasonable move sequences than amateur players. Considering that amateur players originally played some bad moves and that the likelihood approximation was an NN trained using strong human players' games, we presumed it possible that good senses from strong players helped to avoid playing extremely bad moves.

5.3   Selected Move Sequences

For some move sequences where the inferred ones had higher likelihoods and lower territory advantage losses than the actual ones, including both professional and amateur games, we asked Go experts to point out which sequences looked more natural. The experts were not notified which sequences were inferred ones. Generally, the experts considered that the inferred sequences were more natural.

Figure 5 shows 3 pairs where the experts evaluated the inferred sequences as more natural than the actual ones. The top pair was from Pro40m. The likelihood geometric means (abbr. LGMs) for inferred (left) and actual (right) were 0.262

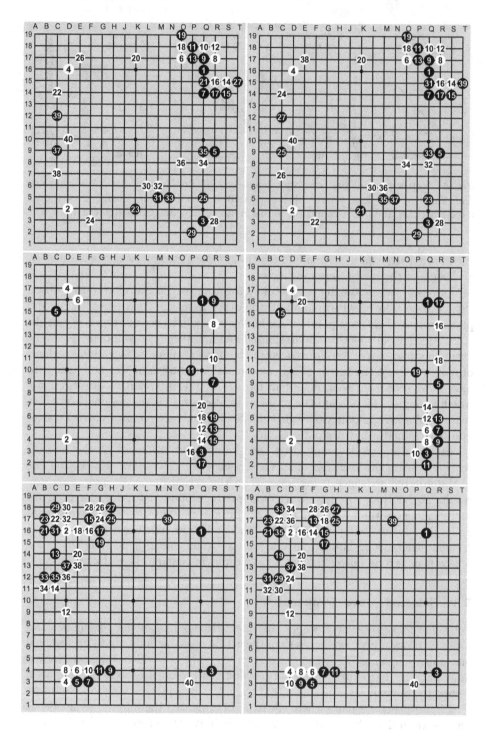

Fig. 5. Example move sequences: inferred on the left and actual on the right.

and 0.206, respectively, and the losses were 0.301 and 0.557. The 1st to the 20th moves were the same. The experts commented that it was natural for Black to play at Q15 to connect the black stones after White played at K17 as the 20th move. This made the shape of black stones thick. KataGo's analysis also showed that Q15 was urgent for both players. In the actual sequence, however, Q15 was omitted until the 31st move. In addition, the experts commented that the timing for Black to play at C9 and C12 to split white stones on the left was more natural after the enclosure of the upper-left corner by E17.

The middle pair was from Ama20m. The LGMs for inferred (left) and actual (right) were 0.153 and 0.084, respectively, and the losses were 0.460 and 0.941. The experts commented that the inferred sequence had more natural stone shapes and the direction of territory development. In contrast, for the actual sequence, the experts commented that the timings of the 15th, 16th, and 20th moves were strange and that the 19th and 20th moves were awful. For the 15th, 16th, and 20th moves, the probabilities from the NN's policy were 0.004, 0.002, and 0.001, respectively. For the 19th and the 20th moves, the losses were 3.673 and 3.150, respectively. Both statistics matched the expert evaluation.

The bottom pair was from Ama40m. The LGMs for inferred (left) and actual (right) were 0.199 and 0.075, respectively, and the losses were 0.814 and 1.292. For the inferred sequence, the experts commented that it was natural for Black to make *atari* (i.e., attempting to capture the opponent's stones) by playing at H17 as the 25th move, though it was a bit strange for Black to attach to the white stone at D3 by playing at E3 (the 5th move). The experts also commented that the flow was natural to make the black stones in the upper-left corner alive. In contrast, for the actual sequence, the experts commented that it was strange for White to make a pole connection by playing at E4 (the 8th move) and for Black to contact E4 from under by playing at E3 (the 9th move). The experts also commented that it was unusual for Black to ignore the cut at G18 from White (the 18th move).

5.4 Ablation Studies

We compared different solution initialization methods and neighbor functions proposed in Sect. 4. Since the results of Pro20m, Pro40m, Ama20m, and Ama40m were similar, we presented those of Pro20m for simplicity. First, regarding solution initialization, we compared the random and greedy methods. When evaluating the solutions by the highest likelihoods within 5 trials, the two methods did not differ too much. However, the stability differed a lot, where the standard deviation of 5 trials' likelihood geometric means was 0.020 for the greedy method but became 0.035 for random. We concluded that the greedy method helped stabilize the likelihood maximization. Second, regarding neighbor functions, we compared swp2, mov2, rot3, and mix. The slopes of the regression lines, like those in Fig. 3, for swp2, mov2, and rot3 were 0.828, 0.912, and 0.686, respectively (mix's was 1.023). When used individually, mov2 performed the best. Mixing different neighbors helped improve the likelihood maximization.

6 Conclusions

Inferring move sequences in Go from stone arrangements on the boards is a skill that many experienced Go players have. We formulated the problem as likelihood maximization and employed a neural network trained using strong human players' games as an approximation of the likelihood function. We adopted simulated annealing, a general optimization algorithm, to solve the maximization problem. We conducted experiments on opening games (20-move and 40-move) played by professional and amateur players. The results showed that the likelihood maximization was well done, though more complex problems (i.e., 40-move cases) still had room to improve. Only a few inferred move sequences exactly matched the actual ones; however, we considered that mismatches were not necessarily bad. We selected some mismatched sequences where the inferred sequences had higher likelihoods and lower territory advantage losses than the actual sequences, and we asked Go experts to evaluate these selected sequence pairs in a blind way. Generally, the experts judged that our approach produced more natural move sequences than those played by humans, including professionals and amateurs.

As interesting applications, our approach can be applied to amateur players as assistant tools. For example, it can be used to infer the move sequences of stone arrangements found in comic books or animations for fun or to infer those found in Go puzzle sets for further study. Regarding future research directions, first, it is interesting to compare move sequences inferred by humans with ours. Second, it is worth improving the scalability of the approach to longer games and more complex settings (e.g., move sequences containing captures). Third, it is promising to try Transformer [9], which is well known to be good at dealing with sequence data and has been applied to a wide variety of tasks.

References

1. Cui, L., Chen, J., He, W., Li, H., Guo, W., Su, Z.: Achieving approximate global optimization of truth inference for crowdsourcing microtasks. Data Sci. and Eng. **6**(3), 294–309 (2021). https://doi.org/10.1007/s41019-021-00164-2
2. Guindon, S., Gascuel, O.: A simple, fast, and accurate algorithm to estimate large phylogenies by maximum likelihood. Systematic Biol. **52**(5), 696–704 (2003). https://doi.org/10.1080/10635150390235520
3. Kang, D.C., Kim, H.J., Jung, K.H.: Automatic extraction of game record from TV Baduk program. In: The 7th International Conference on Advanced Communication Technology (ICACT 2005), pp. 1185–1188 (2005). https://doi.org/10.1109/icact.2005.246173
4. Kirkpatrick, S., Gelatt, C.D., Vecchi, M.P.: Optimization by simulated annealing. Science **220**(4598), 671–680 (1983). https://doi.org/10.1126/science.220.4598.671
5. McIlroy-Young, R., Sen, S., Kleinberg, J., Anderson, A.: Aligning superhuman AI with human behavior. In: Proceedings of the 26th ACM SIGKDD International Conference on Knowledge Discovery & Data Mining, pp. 1677–1687 (2020). https://doi.org/10.1145/3394486.3403219

6. McIlroy-Young, R., Wang, R., Sen, S., Kleinberg, J., Anderson, A.: Learning models of individual behavior in chess. In: Proceedings of the 28th ACM SIGKDD Conference on Knowledge Discovery and Data Mining, pp. 1253–1263 (2022). https://doi.org/10.1145/3534678.3539367

7. Ogawa, T., Hsueh, C.H., Ikeda, K.: Improving the human-likeness of game AI's moves by combining multiple prediction models. In: Proceedings of the 15th International Conference on Agents and Artificial Intelligence (ICAART 2023), pp. 931–939 (2023). https://doi.org/10.5220/0011804200003393

8. Shiba, K., Furuya, T., Nishi, S., Mori, K.: Automatic Go-record system using image processing. IEEJ Trans. Electron. Inf. Syst. **126**(8), 980–989 (2006). https://doi.org/10.1541/ieejeiss.126.980

9. Vaswani, A., et al.: Attention is all you need. In: Advances in Neural Information Processing System (NIPS 2017) (2017)

10. Wu, D.J.: Accelerating self-play learning in Go. In: The 34th AAAI Conference on Artificial Intelligence (AAAI-20). Workshop on Reinforcement Learning in Games (2020). https://arxiv.org/abs/1902.10565

Quantifying Feature Importance of Games and Strategies via Shapley Values

Satoru Fujii[✉]

Kyoto University, Yoshida-honmachi, Kyoto, Japan
`fujii.satoru.75c@st.kyoto-u.ac.jp`

Abstract. Recent advances in game informatics have enabled us to find strong strategies across a diverse range of games. However, these strategies are usually difficult for humans to interpret. On the other hand, research in Explainable Artificial Intelligence (XAI) has seen a notable surge in scholarly activity. Interpreting strong or near-optimal strategies or the game itself can provide valuable insights. In this paper, we propose two methods to quantify the feature importance using Shapley values: one for the game itself and another for individual AIs. We empirically show that our proposed methods yield intuitive explanations that resonate with and augment human understanding.

Keywords: Explainable Artificial Intelligence · Shapley Value · Imperfect Information Game

1 Introduction

Recent advancements in machine learning technology have facilitated the analysis of increasingly large games with greater precision. Notably, the amalgamation of conventional algorithms with deep neural networks—exemplified by AlphaZero [14] in perfect information games, Deep CFR [2] and NFSP [6] in imperfect information games, has demonstrated outstanding performance, occasionally surpassing top human players.

While the development of high-performing AI for games is an intriguing objective, our primary interest lies in deepening our own understanding of the game. In recent years, there has been a rise in professional players endeavoring to enhance their strategies by studying the moves of AI in domains such as shogi. The trend of humans assimilating AI's decision-making processes, or collaborating with AI to reach decisions, is anticipated to proliferate, extending beyond board games to diverse fields. However, these AIs inherently lack transparency. Due to their vastness and intricacy, their strategies elude human comprehension. This absence of explainability poses challenges.

In realms like data analytics and image processing, there is burgeoning interest in eXplainable AI (XAI)—a branch that renders machine learning outcomes interpretable to humans. Yet, in the gaming domain, such research has not been

© The Author(s), under exclusive license to Springer Nature Switzerland AG 2024
M. Hartisch et al. (Eds.): ACG 2023, LNCS 14528, pp. 88–98, 2024.
https://doi.org/10.1007/978-3-031-54968-7_8

sufficiently conducted. At present, human players either unconditionally adopt strategies recommended by advanced AI or resort to observational learning. Introducing explainability to game AI would streamline the process of grasping the AI's thoughts. In addition, explaining AI strategies that are sufficiently close to optimal ones can be considered as characterizing the essence of the game itself, facilitating a profound understanding of the game's nature.

In this paper, we propose two Shapley-based approaches to achieve this goal. Firstly, we introduce a method to quantify the feature importance of the game itself by utilizing the expected return of abstracted games with obscured features. This approach elucidates which features are crucial to act optimally. Secondly, we present a model-agnostic method for calculating the future importance of individual AIs without requiring additional training. This clarifies features that the AI has attention to when choosing actions.

Our empirical findings not only validate their theoretical basis but also reveal that they resonate with and augment human's perceptions of games and strategies.

2 Notation and Background

2.1 Extensive-Form Games

Many games can be formally represented as extensive-form games, comprising the following elements:

1. $\mathcal{N} = (1, \cdots, n)$ denotes the set of players involved in the game.
2. Game tree T is a rooted tree that encapsulates the state transition rules of the game. Z denotes the set of all terminal nodes. Non-terminal nodes are referred to as *turns* and are denoted by X. The outgoing edges originating from a node $x \in X$ are termed *actions* and are denoted as $A(x)$. Since nodes in game tree contain information regarding all preceding actions, these nodes are referred to as the *history*.
3. $\mathcal{P} = (\mathcal{P}_0, \mathcal{P}_1, \cdots, \mathcal{P}_n)$ constitutes a partition of X and signifies the turns for players $1, \cdots, n$. The nodes denoted by $\mathcal{P}_0$ represent *chance* events, where the actions are determined by a probability distribution function $p : \mathcal{P}_0 \to A(\mathcal{P}_0)$.
4. $\mathcal{I}_1, \cdots, \mathcal{I}_n$ form a subpartition of $\mathcal{P}_1, \cdots, \mathcal{P}_n$. Each $I \in \mathcal{I}_i$ is called *infoset*, wherein all nodes in I are indistinguishable to player i. Nodes within the same infoset are required to have an identical set of actions, which can consequently be denoted as $A(I)$.
5. A utility function $u_i : Z \to \mathbb{R}$ yields the payoff for player $i \in \mathcal{N}$ at terminal nodes. If $\sum_{i \in \mathcal{N}} u_i(z) = 0$ for all $z \in Z$, the game is categorized as a *zero-sum* game.

If all infosets of every player consist of a single node, the game is called *perfect information*; otherwise, it is a game of *imperfect information*.

For each player i and for each of their infoset $I \in \mathcal{I}_i$, *behavior strategy* σ_i maps each action $a \in A(I)$ to its selection probability. The set of all possible behavior strategies for player i s denoted by Σ_i.

A strategy profile σ refers to the tuple of behavior strategies for all players. Additionally, σ_{-i} denotes the set of strategies from σ excluding σ_i.

When players act according to a strategy profile σ, the probability of reaching a history h and an infoset I are written as $\pi^\sigma(h)$, $\pi^\sigma(I)$ respectively. The *expected payoff* for player i under strategy profile σ is defined as:

$$u_i(\sigma) = \sum_{h \in Z} \pi^\sigma(h) u_i(h)$$

2.2 Exploitability

Let $b_i(\sigma_{-i})$ denote the best-response payoff of player i against the other players' strategies σ_{-i}, which is formally defined as:

$$b_i(\sigma_{-i}) = \max_{\sigma_i' \in \Sigma_i} u_i(\sigma_i', \sigma_{-i},)$$

Here, $\sigma_i' \in \Sigma_i$ that realizes this maximum is termed as the *best response*. Various methods exist to compute the best-response payoff, such as transforming the problem into a Linear Programming (LP) [13].

A strategy profile $\sigma^* = (\sigma_1^*, \cdots, \sigma_n^*)$ is said to be a *Nash equilibrium* if, for all players $i = 1, \cdots, n$ and for all possible strategies σ_i, the following inequality holds:

$$u_i(\sigma^*) \geq u_i(\sigma_i, \sigma_{-i}^*)$$

It is known that at least one such Nash equilibrium exists.

In two-player zero-sum game, *exploitability* of σ_1, σ_2 is respectively defined as

$$\epsilon_1(\sigma_1) = b_2(\sigma_1) - v_2^*$$
$$\epsilon_2(\sigma_2) = b_1(\sigma_2) - v_1^* \qquad .$$

where v_i^* denotes the expected payoff for player i in a Nash equilibrium.

2.3 Counterfactual Regret Minimization (CFR)

Counterfactual Regret Minimization (CFR) [17] is an iterative algorithm designed to approximate a Nash equilibrium on imperfect information games. During each traversal of the game tree, it computes values termed *counterfactual regrets* for all infosets and subsequently updates its strategy based on the average counterfactual regret accumulated over the traversals.

In two-player zero-sum games, it is established that the average of such strategies converges to a Nash equilibrium. However, the computational complexity of full game tree traversal often renders such methods impractical for large games. As a result, Monte Carlo CFR (MCCFR) [10] is generally preferred, as it samples actions and traverses only a portion of the game tree, thereby providing a more computationally efficient alternative. Deep CFR [2] addresses larger games by using neural networks to approximate average counterfactual regret and average strategy, which eliminates the necessity of storing those values in memory.

2.4   Abstraction

Given the difficulty of applying learning algorithms to intractably large games, *abstraciton* is commonly employed to turn them into smaller-sized games. Formally, an abstraction of infosets α_i is defined as a subpartition of $\mathcal{P}_i$ that is coarser than $\mathcal{I}_i$. Let Γ^{α_i} be the extensive-form game derived by substituting I_i with α_i in the original extensive-form game Γ, and σ_i^{*,α_i} be a strategy of player i in a Nash equilibrim of Γ^{α_i}. It is important to note that infosets of only a single player are abstracted in this case. In two-player zero-sum games, the following theorem has been proven [16]:

Theorem 1. *If abstraction α_i' is a subpartion of α_i,*

$$\epsilon_i(\sigma_i^{*,\alpha_i}) \geq \epsilon_i(\sigma_i^{*,\alpha_i'})$$

This theorem means that an equilibrium strategy in finer abstraction is less exploitable in the original game than one in coarser abstraction, ensuring the theoretical soundness of using abstraction. In addition, it is also pointed out that abstracted player's strategy in this equilibrium is the least exploitable strategy that can be represented in the space [8]. This theorem does not hold when infosets of the other player are also abstracted.

2.5   Feature Importance and Shapley Value

Additive Feature Attribution [12] is a prevalent technique for interpreting machine learning models by assigning importance to each feature of individual inputs. Here, we assume that a model f has an input vector x comprising m features. Further, we assume that f can produce an output even if some of these features are invisible. For $z' \in \{0,1\}^m$, we will use $h_x(z')$ to denote the input derived by obscuring features that have 0 value in z' in the original input x. Let $x' \in \{0,1\}^M$ be the vector that all features visible in x are 1. This implies $h_x(x') = x$.

In additive feature attribution, an *explanation model g*, an interpretable local approximation of the original model around x, is written as:

$$g(z') = \phi_0 + \sum_{i=1}^{m} \phi_i z_i'$$

Here, ϕ_0 is the default output when all features are invisible, and ϕ_i quantifies the contribution of i-th feature to the output value of $f(x)$.

Let $\phi_i(f, x)$ be a function that yields ϕ_j from f and x. Additionally, $z' \setminus j$ represents the vector z' except its j-th element set to 0. The following properties are desired for the explanation model g.

1. Local Accuracy: $f(x) = g(x') = \phi_0 + \sum_{j=1}^{m} \phi_j x_j'$
2. Missingness: $x_j' = 0 \Rightarrow \phi_j = 0$

3. Consistency: For any two models f and f', if

$$f'(h_x(z')) - f'(h_x(z' \setminus j)) \geq f(h_x(z')) - f(h_x(z' \setminus j))$$

for all $z' \in \{0, 1\}^m$, then $\phi_j(f', x) \geq \phi_j(f, x)$.

The following additive feature attribution, known as *Shapley additive explanation* (SHAP) [12], is verified to be the sole additive feature attribution that satisfies all of these properties:

$$\phi_j(f, x) = \sum_{z' \subseteq x'} \frac{|z'|!(m - |z'| - 1)!}{m!} (f(h_x(z')) - f(h_x(z' \setminus j)))$$

In the preceding formula, $|z'|$ is the number of non-zero elements in z, and $z' \subseteq x'$ means z' has non-zero elements that constitute a subset of those in x'. This value is equivalent to the average marginal contributions of feature j across all permutations. Given the challenges of calculating the exact SHAP value, random sampling approaches [15] are often preferred. For neural network models, Deep SHAP [12] offers an efficient computation alternative.

While SHAP focuses on explaining local predictions of the model, there also has been a study using Shapley values for assessing global feature importance, essentially averaging the marginal contributions to guarantee similar desirable properties. [3].

3   Related Work

In the field of Reinforcement Learning (RL) for single agent environments, a significant number of research focused on explainability have been conducted [7]. There is a recent work [1] delving into Shapley-based feature importance in RL using value functions and policies. Another notable contribution [4] employs a saliency map of the game screen, highlighting the agent's focus and effectively indicating the feature importance of each pixel.

For cooperative multi-agent environments, various studies [5,11] have utilized Shapley values to allocate the credit or the reward amongst agents.

4   Descriptions of Our Methods

We introduce two general approaches to quantify feature importance for different subjects: the first seeks to provide a global interpretation of the game itself, while the second aims to elucidate local strategies employed by individual AIs. We focus on two-player zero-sum games. For the target player of explanation i, we postulate that each infoset $I \in \mathcal{I}_i$ can be represented by m features. We let $\mathcal{M}$ be the set of m features. Formally, we assume that feature function $F_j : \mathcal{I}_i \to \mathcal{F}_j$ is given for each $j \in \mathcal{M}$, where each $\mathcal{F}_j$ is a finite set of values.

4.1 Shapley Game Feature Importance

To quantify the global feature importance of the game for the target player i, we compute an approximated Nash equilibrium $\hat{\sigma}^{*,\alpha}$ of games with abstraction α that some of m features are obscured to player i while the opponent plays in null abstraction to hold Theorem 1, then calculate the expected return of i.

For a set of visible features $S \in 2^{\mathcal{M}}$, we define $\alpha_i(S)$ as the abstraction that merges infosets sharing the identical set of actions and feature values $F_j(I)$ for each $j \in S$. The difference between $u_i(\sigma_i^{*,\alpha_i(S)})$ and $u_i(\sigma_i^{*,\alpha_i(\mathcal{M})})$ quantifies the importance of invisible features $\mathcal{M} - S$. This can subsequently be employed to calculate Shapley feature importance by averaging marginal contributions across all possible feature permutations, which we call *Shapley Game Feature Importance (SGFI)*.

4.2 Shapley Strategy Feature Importance

To quantify the feature importance of $\sigma_i(I)$ for a given σ_i, we use the sampled SHAP value [15] with modifications for games, calculated by the following algorithm:

Firstly, we sample $I^{\text{rnd}} \sim \{I' \in \mathcal{I}_i \mid A(I') = A(I)\}$ t_1 times, then assign the average of $\sigma_i(I^{\text{rnd}})$ to ϕ_0. Then, we repeat the following steps t_2 times to calculate ϕ_j for each feature j by averaging the output ϕ'_j.

1. Sample $I^{\text{alt}} \sim \{I' \in \mathcal{I}_i \mid A(I') = A(I)\}$.
2. Randomly choose $\mathcal{O}$ from $m!$ possible permutations of m features.
3. Sample $I^{\text{s}} \sim \{I' \in \mathcal{I}_i \mid F_p(I') = F_p(I),\ F_q(I') = F_q(I^{\text{alt}}),\ \text{for all } p \in P \cup \{j\} \text{ and } q \in Q\}$.
4. Sample $I^{\text{s,alt}} \sim \{I' \in \mathcal{I}_i \mid F_p(I') = F_p(I),\ F_q(I') = F_q(I^{\text{alt}}),\ \text{for all } p \in P \text{ and } q \in Q \cup \{j\}\}$.
5. $\phi'_j \leftarrow \sigma_i(I^{\text{s}}) - \sigma_i(I^{\text{s,alt}})$

In step 3. and 4., P denotes the set of the features ahead of j in $\mathcal{O}$ and Q denotes $\mathcal{M} - P - \{j\}$. If the corresponding infoset does not exist, we assign ϕ_0 for $\sigma_i(I^{\text{s}})$ or $\sigma_i(I^{\text{s,alt}})$. We call $(\phi_1, \cdots, \phi_m)$ computed by this algorithm as *Shapley Strategy Feature Importance (SSFI)*.

5 Experiments

5.1 Goofspiel

We used Goofspiel as the subject of the experiment. The adopted rule set in this study is delineated as follows:

The game is played between two players. Each player starts with a hand of k distinct cards, labeled from 1 to k. Alongside the players' hands, an identical deck of k cards is shuffled to constitute a draw pile.

The gameplay is divided into k rounds. During each round, a single card is randomly drawn from the draw pile and placed face-up at the center of the table.

Subsequently, both players simultaneously select and reveal one card from their respective remaining hands. The player who revealed the card with the higher number receives points equal to the number of the center card. In the case of a tie, neither player is awarded any points. Cards utilized in each round are then excluded from further play. The player accumulating the highest total points across all rounds is declared the winner.

While Goofspiel is designed as a simultaneous-move game, it can be reformulated as a sequential-move game without altering its core mechanics, by allowing one player to select their card first, unbeknownst to the opposing player.

We arranged the following 4 features to explain each infoset of Goofspiel:

1. Center (C) : The center card
2. Deck (D) : Cards remaining in the draw pile
3. Opponent (O) : Cards remaining in the opponent's hand
4. Point (P) : The difference of points between the target player and the opponent

We denote the corresponding feature functions as F_C, F_D, F_O, F_P. This entails $\mathcal{F}_C = \{1, \cdots, k\}$, $\mathcal{F}_D = \mathcal{F}_O = 2^{\{1, \cdots, k\}}$, $\mathcal{F}_P = \{-k(k+1)/2, \cdots, k(k+1)/2\}$. Note that cards remaining in the player's hand constitute the action set and are always visible to the player.

5.2 SGFI of Goofspiel

	SGFI
Center	0.298
Deck	0.295
Opponent	0.096
Point	0.101

Fig. 1. SGFI of Goofspiel

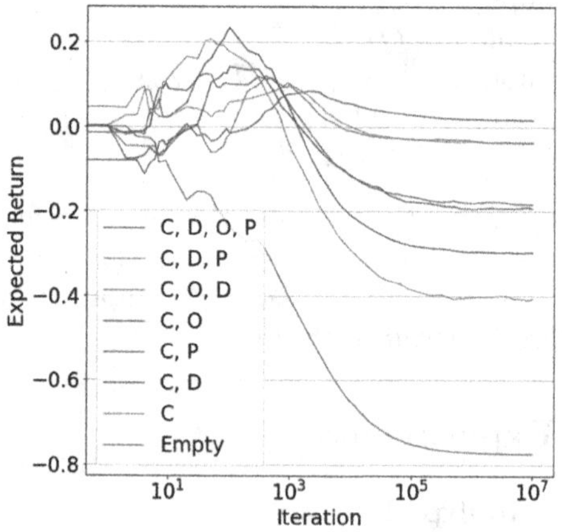

Fig. 2. Expected return convergence of the target player for different abstractions. Legends denote elements in S.

We employed external sampling MCCFR [10] with 10^7 timesteps to approximate a Nash equilibrium strategy for all 2^4 patterns of abstracted Goofspiels in $k = 5$,

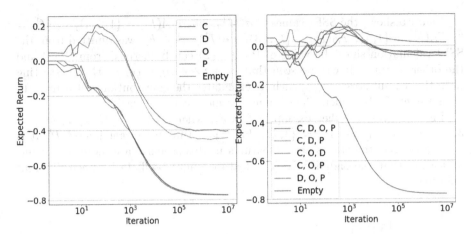

Fig. 3. Convergence of $u_i(\hat{\sigma}_i^{*,\alpha_i(\{j\})})$ and $u_i(\hat{\sigma}_i^{*,\alpha_i(\mathcal{M}-\{j\})})$. Legends denote elements in S.

using OpenSpiel [9] as the learning environment. CFR-BR [8] could be employed for larger games where learning with a null abstraction player is infeasible.

Figure 1 shows the resulting SGFI values of the four features. Given that the most intuitive strategies in Goofspiel involve selecting a high card when a high center card is present and conserving high cards when they remain in the deck, it is intuitive that the feature importances of both the Center and Deck surpass the others.

Figure 2 depicts the convergence of expected return across various abstractions. It can be seen that adding features increases the expected return. It is also noteworthy that when $S = \mathcal{M}(=\{C, D, O, P\})$, the expected return of the target player ended up slightly positive because this abstraction contains sufficient information while a null abstraction suffers slight redundancy, resulting in the target player's faster learning. Figure 3 shows the convergence of $u_i(\hat{\sigma}_i^{*,\alpha_i(\{j\})})$ and $u_i(\hat{\sigma}_i^{*,\alpha_i(\mathcal{M}-\{j\})})$. It can be seen that solely using Opponent or Point does not markedly impact the performance in comparison to $S = \varnothing$, while the combination of these features with the others can significantly influence the outcomes as shown in Fig. 2. It is also noticeable that obscuring a single feature yields similar results regardless of the obscured one. These results underscores the benefits of employing the Shapley value as an index instead of simply using $u_i(\hat{\sigma}_i^{*,\alpha_i(\{j\})}) - u_i(\hat{\sigma}_i^{*,\alpha_i(\varnothing)})$ or $u_i(\hat{\sigma}_i^{*,\alpha_i(\mathcal{M})}) - u_i(\hat{\sigma}_i^{*,\alpha_i(\mathcal{M}-\{j\})})$ as the feature importance of $j \in \mathcal{M}$.

5.3 SSFI of Goofspiel AI

We once again utilized external sampling MCCFR with 10^6 timesteps for Goofspiel in $k = 4$ without any abstraction in order to obtain the AI to explain. This resulted in strategy profile $\hat{\sigma}*$ exhibiting an average exploitability of 0.006. In SSFI calculations, both t_1 and t_2 were set to 10^6 (Figs. 4 and 6).

Figure 5 shows the SSFI values of $\hat{\sigma}_1^*(I^1)$, where $A(I^1) = \{1,2,4\}$, $F_C(I^1) = 3$, $F_D(I^1) = \{1,4\}$, $F_O(I^1) = \{1,2,3\}$, $F_P(I^1) = -2$. F_P was omitted from the SSFI calculation since the value of Point (-2) can be derived from $A(I^1)$ and the other features. It can be seen that $\phi_0 + \phi_C + \phi_D + \phi_O = \hat{\sigma}_1^*(I^1)$: starting from the default strategy ϕ_0, each ϕ_j elucidates the contribution of feature j to the strategy. This renders the explanation more comprehensible.

Figure 7 gives another example of SSFI, where $A(I^2) = \{1,4\}, F_C(I^2) = 3$, $F_D(I^2) = \{4\}$, $F_O(I^2) = \{3,4\}$, $F_P(I^2) = 3$ in this target infoset I^2. It is noticeable that the values across features in this scenario are more congruent, indicating that the significance of a feature might oscillate between different infosets.

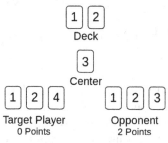

Fig. 4. Situation in I^1

	Card 1	Card 2	Card 4
ϕ_0	32.5%	35.3%	32.2%
ϕ_C	-16.3%	+16.1%	+0.2%
ϕ_D	-9.6%	+16.9%	-7.3%
ϕ_O	-6.5%	+12.7%	-6.2%
$\hat{\sigma}_*(I^1)$	0.1%	81.0%	18.9%

Fig. 5. SSFI of $\hat{\sigma}_1^*(I^1)$.

Fig. 6. Situation in I^2

	Card 1	Card 4
ϕ_0	49.5%	50.5%
ϕ_C	+5.4%	-5.4%
ϕ_D	+4.5%	-4.5%
ϕ_O	+3.8%	-3.8%
ϕ_P	+4.2%	-4.2%
$\hat{\sigma}_*(I^2)$	67.0%	33.0%

Fig. 7. SSFI of $\hat{\sigma}_1^*(I^2)$.

6   Conclusion

In this paper, we have put forward two methods for the quantification of feature importance, each with its distinct application: understanding the game on a macro level and deciphering the attention of individual AI strategies. Other than providing the theoretical backgrounds of those methods, we have empirically demonstrated that our methods can explain games and strategies in a way that aligns with and complements human intuition.

We need to keep in mind that XAIs using feature importance have inherent limitations in that they solely teach you what feature they regard as important. However, we posit that such feature importance can enhance our understanding of more intricate strategies and provide better heuristics for the game.

We anticipate that interpretative methodologies, such as those we have proposed, will enhance human learning supported by AI and streamline collaborative decision-making processes between humans and AI.

Acknowledgement. I would like to thank Professor Hideki Tsuiki for meaningful discussions. I am also grateful to the referees for useful comments.

References

1. Beechey, D., Smith, T.M., Şimşek, Ö.: Explaining reinforcement learning with Shapley values. In: International Conference on Machine Learning, pp. 2003–2014. PMLR (2023)
2. Brown, N., Lerer, A., Gross, S., Sandholm, T.: Deep counterfactual regret minimization. In: International Conference on Machine Learning, pp. 793–802. PMLR (2019)
3. Covert, I., Lundberg, S.M., Lee, S.I.: Understanding global feature contributions with additive importance measures. Adv. Neural. Inf. Process. Syst. **33**, 17212–17223 (2020)
4. Greydanus, S., Koul, A., Dodge, J., Fern, A.: Visualizing and understanding Atari agents. In: International Conference on Machine Learning, pp. 1792–1801. PMLR (2018)
5. Han, S., Wang, H., Su, S., Shi, Y., Miao, F.: Stable and efficient Shapley value-based reward reallocation for multi-agent reinforcement learning of autonomous vehicles. In: 2022 International Conference on Robotics and Automation (ICRA), pp. 8765–8771. IEEE (2022)
6. Heinrich, J., Silver, D.: Deep reinforcement learning from self-play in imperfect-information games. arXiv preprint arXiv:1603.01121 (2016)
7. Heuillet, A., Couthouis, F., Díaz-Rodríguez, N.: Explainability in deep reinforcement learning. Knowl.-Based Syst. **214**, 106685 (2021)
8. Johanson, M., Bard, N., Burch, N., Bowling, M.: Finding optimal abstract strategies in extensive-form games. In: Proceedings of the AAAI Conference on Artificial Intelligence, vol. 26 (2012)
9. Lanctot, M., et al.: OpenSpiel: a framework for reinforcement learning in games. CoRR **abs/1908.09453** (2019). http://arxiv.org/abs/1908.09453
10. Lanctot, M., Waugh, K., Zinkevich, M., Bowling, M.: Monte Carlo sampling for regret minimization in extensive games. Adv. Neural. Inf. Process. Syst. **22**, 1078–1086 (2009)
11. Li, J., Kuang, K., Wang, B., Liu, F., Chen, L., Wu, F., Xiao, J.: Shapley counterfactual credits for multi-agent reinforcement learning. In: Proceedings of the 27th ACM SIGKDD Conference on Knowledge Discovery & Data Mining, pp. 934–942 (2021)

12. Lundberg, S.M., Lee, S.I.: A unified approach to interpreting model predictions. In: Advances in Neural Information Processing Systems, vol. 30 (2017)
13. Nisan, N., Roughgarden, T., Tardos, E., Vazirani, V.V.: Algorithmic Game Theory. Cambridge University Press, USA (2007)
14. Silver, D., et al.: A general reinforcement learning algorithm that masters chess, shogi, and go through self-play. Science **362**(6419), 1140–1144 (2018)
15. Štrumbelj, E., Kononenko, I.: Explaining prediction models and individual predictions with feature contributions. Knowl. Inf. Syst. **41**(3), 647–665 (2014)
16. Waugh, K., Schnizlein, D., Bowling, M.H., Szafron, D.: Abstraction pathologies in extensive games. In: AAMAS (2), pp. 781–788 (2009)
17. Zinkevich, M., Johanson, M., Bowling, M., Piccione, C.: Regret minimization in games with incomplete information. Adv. Neural. Inf. Process. Syst. **20**, 1729–1736 (2007)

Player Investigation

The Impact of Wind Simulation on Perceived Realism of Players

Zeynep Burcu Kaya Alpan[1,2]([✉]) [iD] and Şenol Pişkin[3] [iD]

[1] Communication Sciences PhD Programme, Istinye University, Vadi Campus,
Sariyer, Istanbul, Turkey
[2] Digital Game Design Department, Bahcesehir University, Galata Campus,
Karakoy, Istanbul, Turkey
zeynepburcu.kayaalpan@stu.istinye.edu.tr
[3] Department of Mechanical Engineering, Faculty of Engineering and Natural
Sciences, Istinye University, Vadi Campus, Sariyer, Istanbul, Turkey
senol.piskin@istinye.edu.tr

Abstract. This study aims to investigate the impact of wind to players'
perceived realism in digital games. In digital games, there is an ongoing
and never ending quest for realism. As games keep getting more and more
graphically demanding, game designers try to include senses in order to
invoke realism. Perceived game realism is a multidimensional construct
and while much attention has been given to visual fidelity, there are subtle elements in physical reality, such as wind blowing and floating dust,
that contribute to the perception of realism.

Previous studies have examined general aspects of realism in various
genres of games, however the specific impact of wind in digital games
remains largely unexplored. This study aims to address this gap by integrating real-time, dynamic wind simulation into a game environment and
explore changes to the perceived realism of players. The research focuses
on finding out which aspects of perceived realism are affected by adding
dynamically simulated wind within a third-person shooter game context.

The findings of this study contribute to understanding how wind simulation impacts perceived realism in digital games. The insights gained
can inform game developers and designers in creating more immersive
and realistically perceived game environments. By investigating the role
of environmental wind in enhancing the sense of realism, this research
advances our understanding of the multi-faceted nature of player engagement and the importance of incorporating multi-modal, realistically simulated elements in game design.

Keywords: Wind simulation · Perceived realism · Digital games

1 Introduction

In the gaming industry's relentless pursuit of realism, there has been a pre-
dominant emphasis on graphical fidelity. However, perceived game realism is a

© The Author(s), under exclusive license to Springer Nature Switzerland AG 2024
M. Hartisch et al. (Eds.): ACG 2023, LNCS 14528, pp. 101–110, 2024.
https://doi.org/10.1007/978-3-031-54968-7_9

multidimensional construct (Ribbens et al., 2016a; 2016b). Beyond high visual fidelity, elements like narrative, interaction, character actions, embodiment, and environmental sounds contribute to the perception of game realism.

Previous research on perceived realism in games has examined various dimensions, including social realism (Galloway, 2004), the impact of controllers on perceived realism (McGloin et al., 2011), and the connection between perceived realism and player enjoyment in historical games (Vandewalle et al., 2023).

Much of this research has focused on shooter games (Ribbens and Malliet, 2010; Ribbens et al., 2016a; 2016b; Daneels et al., 2018). This focus is attributed to the genre's popularity and its often violent nature. Because realistic behaviors in games can encompass realistic depictions of violence, there is an emphasis on measuring perceived realism of shooter games. However, research shows that individuals who enjoy violent behaviors in games do not necessarily exhibit such behaviors in the real world. To understand this dichotomy, it is crucial to examine perceived realism, which enhances the player's sense of presence and enjoyment in the gaming world (Daneels et al., 2018).

Quantitative studies on perceived realism in games (Ribbens, 2013a; Ribbens, 2013b; Ribbens and Malliet, 2010; Ribbens et al., 2016a; 2016b; Vandewalle et al., 2023) have resulted in the development of a scale validated for shooter games (Ribbens et al., 2016a; 2016b).

However, these studies have not specifically investigated the impact of wind on realism. This study aims to address this gap by examining how realistic and dynamically simulated wind affects perceived realism.

One way to increase perceived realism is adding realistic flow simulation such as wind. Wind is an important element our physical reality that is sensed with haptic, auditory and visual senses. Floating dust, increasing and decreasing sound of howling wind, flutter of tree leaves and grass, swinging of flags in the sky are perceived through senses, meaning wind to the human brain.

1.1 Wind Simulation in Games

Wind is one of the elements that contribute to physically perceived realism in digital games. In games striving for realism, wind adds subtle details, such as swaying vegetation, billowing fabrics, drifting snow, and floating dust, enhancing the perception of realism. These features demand high-performance capabilities from computer systems.

Wind simulation can be incorporated in the design of a game in two ways. On the one hand there are games which have wind as a background element, a distant feature of the environment. On the other hand several games incorporate wind as a main design element. Games such as The Legend of Zelda: The Wind Waker (2002) and PlayerUnknown's Battlegrounds (2017) use wind as a design element that changes conditions of the game, whereas Ghost of Tsushima uses it as a way to guide the player. Simulation games like Microsoft's Flight Simulator replicate weather conditions to realistically control air vehicles.

Another way to incorporate wind in games is to use physical wind simulation mechanisms. Arcade games that feature motorcycle racing use fans to simulate the impact of wind on the driver.

Sound design is an important element of games because when visual and auditory senses work together, the player may perceive it to be more realistic. For example in Wind Waker, the sound of wind disappears rapidly as the player character enters a cave. The lack of wind sound draws attention to the environmental condition that is now gone. Change of environment becomes more apparent.

On the other side, player communities also increasingly express interest in adding such realism-enhancing features themselves, often through self-developed "mods." Technological advancements are also raising the possibility of making such simulations a permanent part of games or allowing players to toggle wind effects on and off according to their preferences. Yet, the impact of dynamically simulated wind on perceived realism remains unexplored from a scientific perspective.

2 The Game

This section describes the design of the game that was developed for the study. The genre was chosen to be a third person shooter game because the scale was mostly validated in this genre of games. Specifically, a sci-fi setting was used. The wind system used in the game is dynamic, which means that several parameters change randomly to simulate changes in the wind condition. Cloth simulation, particle effects material effects and sound effects were used in synchrony to mimic dynamically changing weather conditions.

2.1 Game Design

The game was developed using Unreal Engine 5.0 by Epic Games. This game engine features a sample game project called "Lyra". Lyra is a modular shooter game template that can be used to develop different games through its different game modes. It features several different level designs, comprehensive enemy AI, multiplayer capabilities and a flexible UI system that is ready to use out of the box. Lyra template was used to develop a game with easy gameplay and typical controls, for single player use. This new game was named "Rewind" (see Fig. 1).

The AI of the game creates bots for both the enemies and the teammate of the player. However the AI system was found to be working too efficiently, so it was deliberately slowed down for the sake of the player experience. Multiplayer system and lobby level was deactivated because the game is going to be used with its AI system only.

Flags that are useful for guiding players were simulated with cloth simulation. Foliage such as weeds and flowers and also tree leaves were moved through material effects. Particle effect depicting wind through steam sources around control zones were implemented.

Fig. 1. Logo of the game, featuring the game's name Rewind.

The game mode called "control" inside the Lyra starter project was chosen as the main gameplay design. In this game mode, win condition is such: Whichever team gets to 120 points through securing control zones wins. The team obtaining at least 2 zones out of 3 starts gaining 1 point per second. This means upon obtaining control over most zones at the start of the game, a team may theoretically finish it without ever losing them, because the teams consist of 2 players. In the study, participants play together with an AI controlled bot, similarly skilled to the AI controlled enemy bots. Therefore, a single play session might take from 120 to 240 seconds.

3   Study

This research aims to measure the influence of realistically and dynamically simulated wind on the perceived realism of university students within the context of a third-person shooter game. The study involves integrating wind simulation that dynamically affects game elements like fabric and vegetation, as well as providing an environmental wind sound effect. A generic third-person shooter game in a sci-fi setting was developed for this purpose (see Figs. 2 and 3).

During testing, wind simulation is active or inactive during gameplay sessions with participants. An evaluation is conducted through surveys in order to identify changes in perceived realism among players.

- RQ1 Does dynamically simulated wind impact players' perceived realism?
  If players experience higher levels of perceived realism in their gameplay with simulated wind effects compared to their game session without simulated wind, the results will show that the items of the scale are impacted by wind simulation.
- RQ2 Are certain items of the scale impacted more by the wind simulation? If so, which items?

Fig. 2. The opening menu and waiting scenes of the game are demonstrated on top of each other.

Fig. 3. A typical gameplay screen from when the game starts. Players own team is red. Trees, flags and smoke effect as well as the gameplay UI are present in the image.

The scale used in the study, called Perceived Realism in Games, has 32 items related to different effects of the game such as agency (freedom of choice), player behaviour (simulational realism), visual fidelity and sound effects (perceptual pervasiveness), player experience (involvement) and the role of the developers (authenticity).

3.1   Method

University students in bachelor degree, lab visitors and high school graduates over the age of 18 have been invited to the study. After being informed about the study and signing consent forms, participants firstly played the tutorial level of the game. In the tutorial level, wind-related game elements were excluded. After playing for a single round or until they are confident in their abilities in the game for minimum 2 min, participants started playing either version "A" or version "B" by selecting it from the menu, as they are directed by the research team.

Participants who select "A" first, play the level for at least 3 min and proceed to the questionnaire. After completing the online questionnaire by marking "A" on a different computer for about 15 min, they play the game version "B" and complete the questionnaire for a second time. Half of the participants were guided to start with "A" and proceed to "B", and the other half did it in the opposite way.

In total, 35 participants completed both questionnaires successfully, out of which 13 are female and 22 are male.

3.2   Result

According to the results, the age of the participants ranged from 18 to 41 years, with a mean age of 23.69 years (SD = 6.22). The descriptive statistics indicate a diverse age range among the 35 participants, contributing to the variability observed in the study sample.

Players have described their gameplay habits with a 5 point Likert scale like other items, which resulted in the mean and median results in Table 1. The mean score for playing shooting games falls between "monthly" (3) and "weekly" (4), indicating that, on average, the surveyed players play shooting games around once a month to once a week. The median score of 3.00 corresponds exactly to "monthly," suggesting that the most common response among the players is to play shooting games on a monthly basis. This indicates that a significant portion of players engage in shooting games periodically, with some playing less often and some playing more frequently Table 1).

The mean score for playing digital games in general falls between "weekly" (4) and "daily" (5), indicating that, on average, the surveyed players play digital games in general approximately once a week to almost daily. The median score of 4.00 corresponds to "weekly," suggesting that the most common response among the players is to play digital games in general on a weekly basis. This indicates

Table 1. Player Habits

	N	Minimum	Maximum	Mean	Std. Deviation
How often do you currently play shooting games?	35	1	5	3.06	1.514
How often do you currently play digital games in general?	35	1	5	3.46	1.502

that the players have a habit of engaging in digital games regularly, with some playing less often and some playing almost daily.

Based on these results, the surveyed players tend to play digital games in general more frequently than they play specifically shooting games. On average, they play shooting games around once a month to once a week, while they play digital games in general approximately once a week to almost daily. This suggests that the players have a habit of playing digital games regularly, whereas shooting games somewhat less frequently.

Overall means for the survey done after wind enabled and the survey after wind disabled were computed. A simple comparison between the means of wind enabled and wind disabled mean scores indicate that, on average, participants rated the wind enabled version slightly higher than the wind disabled version (Table 2).

However, the standard deviations suggest that there is some variability in responses, and the difference between the versions may not be uniform across all participants. Overall mean scores were also compared using Wilcoxon Signed Ranks Test, which is a non-parametric, paired samples test. Comparing the overall mean scores between two conditions showed significant difference (n=35, p=0.009). Figure 1 shows the difference between wind disabled (B) and enabled (A).

Table 2. Overall Means for Wind Enabled and Disabled

	N	Minimum	Maximum	Mean	Std. Deviation
Wind Enabled	35	2.03	4.61	3.3719	.65361
Wind Disabled	35	1.65	4.77	3.2410	.76166

Individuals items were also compared using the Wilcoxon Signed Ranks Test. Because the sample size is n=35, normal distribution was assumed. The following items show results of this analysis revealed that the following 4 items show significant difference with significance level of 0.05 :

1. By playing this game, I can learn how to behave in real life. (p=0.033)
2. I felt I determined the course of the game. (p=0.049)
3. The sound effects in the game were impressive. (p=0.011)

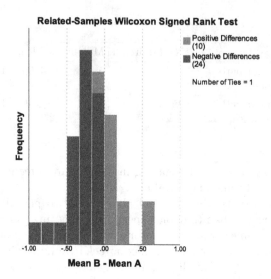

Fig. 4. Related-Samples Wilcoxon Signed Rank Test

4. The developers of the game have examined which objects (e.g. weapons) fit within the context of the game. (p=0.005).

Other items do not show statistically significant differences.

3.3   Discussion

The sample size of the study is rather small yet show some meaningful statistically significant results. However it is important to acknowledge the limitations of the current study in fully explaining the implications of wind simulation. Therefore this research findings should be read as preliminary findings and more research should be conducted to show more meaningful results. Overall mean scores as well as certain items of the scale show meaningful difference between wind enabled and wind disabled version. They do not correlate to the 6 dimensions of the scale (freedom of choice, simulational realism, perceptual pervasiveness, involvement and authenticity). Instead, all 4 items are from different dimensions.

The findings of this study support RQ1, revealing a statistically significant difference in overall mean scores between the wind-enabled and wind-disabled versions. This indicates that players, on average, experience varying levels of perceived realism influenced by the presence or absence of simulated wind effects. These results align with the hypothesis that the items of the scale are impacted by wind simulation.

However, while the overall means demonstrate a significant difference, further analysis is required to discern the specific dimensions or items within the scale that are most affected by the introduction of simulated wind. This leads to RQ2, which tries to identify specific items within the scale that are more susceptible to the impact of wind simulation. The results indicate that certain items each from different dimensions exhibit meaningful differences between the wind enabled and wind disabled versions. This could imply that the influence of simulated wind is nuanced and doesn't uniformly affect all aspects of player experience.

4   Conclusion

While the pursuit of graphical realism in games continues, this research highlights that perceived game realism encompasses multiple dimensions. While visual fidelity has received significant attention, there are subtle yet crucial elements from the physical world, like the sensation of wind simulated through distant yet dynamic sound effects, particle effects such as floating dust that changes direction every so often, and textiles that float about, that contribute to the overall sense of realism.

Despite previous research exploring various aspects of realism in games, the specific influence of wind in digital games has largely remained unexplored. This study is a first step to bridging this gap by introducing real-time wind simulation into gameplay and examining how this affects the perceived realism. However the small sample size and the difficult genre chosen for the game should be taken into consideration. The third person shooter with control zone gameplay mode was somewhat difficult to play for participants who are unfamiliar with the game. On the other hand, participants who are shooter-game-savvy may have negatively impacted by the choice of game genre because they were heavily concentrated on their success in the game.

The research primarily focuses on how players experience enhanced realism through simulated wind effects in the context of a third-person shooter game. However, to understand the impact of wind simulation in games, the impact of physical wind simulated through mechanical systems in synchrony with the game should be explored. A second study that includes a mechanical structure that physically simulates wind is being devised. This game will be simpler in order not to impact findings.

Another way to interpret this study is to say that minor environmental elements do not impact the overall game experience too much. Visual and auditory environmental elements are rather undetected by the player, who is concentrated on the gameplay and decision making process. For the sake of the research design, the game could not feature wind as a main design element, because otherwise the control version of the game without the wind would not work. But a game designer who wants to use such game elements in the design would make it an indispensable part of the game. This would make the simulated wind elements much more noticeable for the player.

Through incorporating these simulated wind effects, the study emphasises the importance of environmental game elements. With the ongoing technical advances, realistic and dynamic physics simulations may become more mainstream in game technologies. The study also shows the importance of game design choices and to be mindful and decisive in implementing game elements. Because if the player's attention is not drawn to it, it may be rather unnecessary and may be excluded from the game. In order to draw attention to the wind, perhaps the game should also be centered around wind. In which case, sound effects and game related objects such as weapons might be especially important to consider according to the results of the study.

Acknowledgments. This research received partial funding support from the European Research Executive Agency Marie-Sklodowska Curie Actions - Individual Grant (101038096), TUBITAK ARDEB 3501 (121E732) and Istinye University Scientific Research Projects projects (2019B1).

References

Daneels, R., Malliet, S., Koeman, J., Ribbens, W.: The enjoyment of shooting games: Exploring the role of perceived realism. Comput. Hum. Behav. **86**, 330–336 (2018). https://doi.org/10.1016/j.chb.2018.04.053

Galloway A. Social realism in gaming. Game Studies, 4. (2004). http://www.gamestudies.org/0401/galloway/

Ghost of Tsushima. (2020). Sony Interactive Entertainment

McGloin, R., Farrar, K., Krcmar, M.: The impact of controller naturalness on spatial presence, gamer enjoyment, and perceived realism in a tennis simulation video game. Presence: Teleoperators Virtual Environ. **20**(4), 309–324 (2011)

Microsoft's Flight Simulator (2020). Microsoft

PlayerUnknown's Battlegrounds (2017). Krafton

Ribbens, W.: In search of the player: Perceived realism and playing styles in digital game effects [unpublished doctoral dissertation]. Leuven: KU Leuven (2013a). https://lirias.kuleuven.be/handle/123456789/414856

Ribbens, W.: Perceived game realism: a test of three alternative models. Cyberpsychol. Behav. Soc. Netw. **16**(1), 31–36 (2013)

Ribbens, W., Malliet, S. Perceived digital game realism: a quantitative exploration of its structure. Presence: Teleoperators and Virtual Environments, **19**(6), 585–600 (2010)

Ribbens, W., Malliet, S., Van Eck, R., Larkin, D.: Perceived realism in shooting games: Towards scale validation. Comput. Human Behav. **64**, 308–318. (2016) https://doi.org/10.1016/j.chb.2016.06.055

Ribbens, W., Malliet, S., Van Eck, R., Larkin, D.: Perceived game realism scale [Database record]. APA PsycTests (2016). https://doi.org/10.1037/t57201-000

The Legend of Zelda: The Wind Waker (2002). Nintendo

Vandewalle, A., Daneels, R., Simons, E., Malliet, S.: Enjoying my time in the animus: a quantitative survey on perceived realism and enjoyment of historical video games. Games Culture **18**(5), 643–663 (2023)

Hades Again and Again: A Study on Frustration Tolerance, Physiology and Player Experience

Maj Frost Jensen, Laurits Dixen, and Paolo Burelli[✉][iD]

Center for Digital Play, IT University of Copenhagen, Copenhagen, Denmark
{mfje,ldix,pabu}@itu.dk

Abstract. Accurately quantifying player experience is challenging for many reasons: identifying a ground truth and building validated and reliable scales are both challenging tasks; on top of that, empirical results are often moderated by individual factors. In this article, we present a study on the rogue-like game Hades designed to investigate the impact of individual differences in the operationalisation of player experience by cross-referencing multiple modalities (i.e., questionnaires, gameplay, and heart rate) and identifying the interplay between their scales.

Keywords: player experience · game experience questionnaire · behavioural inhibition system · behavioural activation system · heart rate · rogue-like

1 Introduction

Player experience modelling plays a crucial role in game development and game research by providing insights into how players perceive, interpret, and engage with games. It focuses on understanding gameplay's psychological, emotional, and cognitive aspects, allowing game designers and researchers to create more immersive and enjoyable experiences for players [20].

However, accurately quantifying different facets of player experience remains a challenging problem to this day for many reasons. Player experience is inherently subjective and is influenced by various contextual factors, such as individual preferences, cultural background, and previous gaming experiences [18]. Furthermore, it comprises multiple dimensions, including emotions, cognitive engagement, social interaction, and immersion. These dimensions interact with each other and are also influenced by personal differences [23].

Finally, selecting appropriate tools and methods to quantify the experience of the players is challenging. Traditional survey-based methods may not effectively capture the nuances of subjective experiences and are summative in nature. Researchers need to develop innovative techniques, such as physiological measures (e.g., heart rate variability), behavioural observation, or eye tracking, to

© The Author(s), under exclusive license to Springer Nature Switzerland AG 2024
M. Hartisch et al. (Eds.): ACG 2023, LNCS 14528, pp. 111–120, 2024.
https://doi.org/10.1007/978-3-031-54968-7_10

complement self-report measures and provide a more holistic understanding of the player experience [22].

Most dimensions of the player experience, such as engagement, enjoyment, or frustration, all share these common challenges. For instance, some people have a high frustration tolerance and are less likely to get as annoyed or frustrated at minor setbacks, whereas those with a low frustration tolerance easily grow agitated at the same inconveniences. These individual differences are likely to have an impact on the operationalisation of frustration.

Within player experience research, these individual and contextual differences are often overlooked, as they rely on information that is often intangible and external to the games; however, aspects such as frustration tolerance have been studied extensively in other fields [12,15].

In this paper, we present a study that aims to investigate the moderation effect of contextual and individual differences on the player experience. In the study, we analyse and evaluate the player experience of a sample of players playing Hades [8], a competitive rogue-like action game, and investigate the interplay between a number of self-reported scales, players' performance, their heart rate, and their tolerance to stress and frustration.

With this study, we intend to shed some light on the following research questions:

Q1 Do individual differences in frustration tolerance have a measurable impact on the player experience?
Q2 Do individual differences in the psychophysiological responses of players have a measurable relationship with the player experience?

Although the study is not intended to give a comprehensive answer to either of the two questions given the limitations of its scope and design; we selected the game and designed the study to maximise the likelihood of finding some evidence of these relationships, if they exist.

2   Related Work

Frustration and how it relates to gaming has been studied in many different contexts, this includes studies into how near-misses, despite causing frustration, can increase the urge for players to continue playing Candy Crush [10], adaptive design in video games where frustration plays a role [24], as well as dynamic difficulty adjustment [25].

One of the primary challenges with employing frustration or other aspects of the player experience for adaptation of dynamic difficulty adjustment is the operationalisation of the chosen construct. Several studies have attempted to quantify frustration; among these, probably the most common approach is through questionnaires, with instruments such as the Game Experience Questionnaire (GEQ) [9], Game User Experience Satisfaction Scale (GUESS) [14] or ENJOY [6].

These questionnaires offer powerful and validated tools to estimate several aspects of player experience; however, they heavily rely on players' self-awareness and offer low granularity in terms of which aspect of the game experience and the player behaviour has elicited a specific player reaction.

To alleviate these limitations, several researchers have investigated how to operationalise player experience through other means. For example, Shaker et al. [16] investigate the fusing of gameplay data and head movement, while Burelli et al. [1] used full-body posture and movement. Other researchers have investigated the use of psychophysiological signals, such as heart rate [7] and galvanic skin response [5], showing how they can be powerful game-independent markers of player experience [13].

While some of the aforementioned studies attempt to infer individual differences from data collected within the analysed gaming experience, most of them do not take into account contextual individual differences in terms of personality and attitude, which has been identified as a potential factor by Canossa et al. [2] and Chang et al. [4]. In this paper, we attempt to further explore this dimension of the player experience, estimate the impact of these contextual differences, and evaluate whether they can be measured and detected.

3   Methods and Materials

To estimate the interplay between personal attitude, reported player experience, and player behaviour, we conducted an empirical study of player experience in the game Hades [8]. During the experiment, we collected self-reported feedback on the player experience using GEQ. Only the 'Core' and the 'Post-game' modules were used as the modules 'Social Presence' and 'In-Game' were not relevant to the experimental setup, as the game was played alone and we did not ask participants to answer questions during the gameplay. Instead, we collected in-game player behaviour and the players' heart rate through the Polar H10 ECG sensor[1]. The last aspect captured is the players' tolerance to frustration, estimated using the Behavioral Inhibition/Behavioral Activation Scale (BIS/BAS) [3] questionnaire, paired with the Frustrative Nonreward Responsiveness (FNR) [21] questionnaire.

The BIS/BAS scales are primarily used to assess dispositions to anxiety, depression and other mental disorders, but have also proven to be helpful in finding relations in personality traits [19]. The questionnaire is structured into 4 sub-scales; BIS, BAS Drive, BAS Fun Seeking, and BAS Reward Responsiveness. BIS explains the tendency of people to avoid negative outcomes; this means that high BIS scores in individuals are related to higher feelings of anxiety. BAS and its subscales explain individuals' tendency to respond to rewards or to engage in goals where the possibility of reward is there. FNR measures the motivational response to the lack of reward, acting as an extended subscale to BIS/BAS. Both scales are Likert scales with respectively 4 or 5 levels.

---

[1] https://www.polar.com/uk-en/sensors/h10-heart-rate-sensor.

The study includes 20 participants with a mean age of 27, mostly consisting of students from the IT University of Copenhagen and their acquaintances. Before analysis, each gameplay video has been coded to identify the game events, detailed in the following sections. The collected data, the video coding, and the scripts used to analyse the data can be downloaded at https://github.com/itubrainlab/hades_player_experience/.

3.1   The Game

Hades is a single-player, rouge-like game in which the player proceeds through a series of progressively more challenging rooms, fighting enemies and collecting boons (Fig. 1 Top). Death results in a total reset of the progress both in rooms and power-ups (boons) the player had accumulated during a run, and a return the initial stage (Fig. 1 Bottom). In this way, the game is widely viewed as a challenging and potentially frustrating game to play, as any small mistake could cost the player a lot of progress.

We included only a single game, and any finding must be interpreted with this in mind. However, that also allows game-specific features to be extracted, which hopefully gives a more detailed understanding compared to general performance measures.

3.2   Protocol

Each recording session is individual and the player plays the first few levels of Hades for approximately 30 min using a joy-pad controller. All sessions were conducted according to the following procedure:

1. Setup the Polar H10 and make sure it connects with the Polar Flow app.
2. Setup Hades and the controller.
3. Test-run the recording software to make sure it works.
4. Start a new save/profile on Hades on Hell mode.
5. Invite the participant in and make sure they are comfortable.
6. Ask them to put on the heart-rate monitor. Have a picture ready to show how it's supposed to sit. Give them access to a bathroom and a towel so they can put it on in privacy.
7. Have the participant sit down and make sure they are comfortable while connecting the sensor to the app.
8. Let the player know that the first chamber will not contain enemies so they can run around and get used to the controller before continuing.
9. Let the player know that there will be a menu at the start of each new run but all they need to do is click 'Begin Escape' when it pops up.
10. Tell the player that once the 30 min are up, you will place a paper next to them stating "Last Run!" which means they continue to play until they complete the current, and then the Play Session will be over.
11. Start the screen-recording software at the same time as the app.

Fig. 1. Screenshots of the Hades. Top: example of combat with multiple enemies, player is about to lose. Bottom: after losing the player character returns to the beginning again, losing almost all progress from the previous run.

12. Sit for 1–3 min to capture the normal heart rate and then tell them they can begin playing.
13. After the 30 min are up, discreetly place the piece of paper that says "Last Run!" next to the player, and wait for them to finish the current run.
14. Let them know that they can remove the heart rate monitor and make sure that they are ready to complete the closing questionnaire.
15. Bring up GEQ on the computer and let them fill it out.

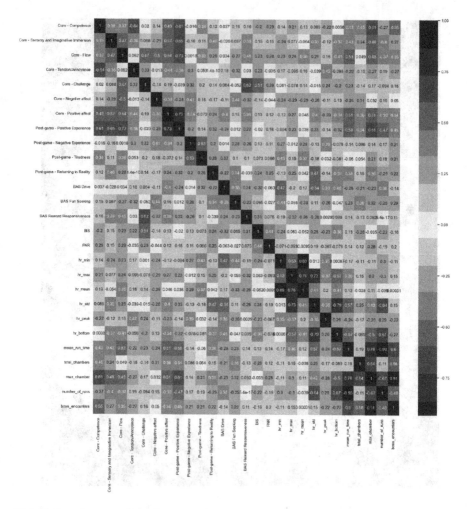

Fig. 2. Correlation matrix (Pearson's) between the scales from the two questionnaires and the features extracted from the heart rate and game events.

4 Results

4.1 Features Extraction

The heart rate data recorded is structured as a time series of heart rates recorded at a frequency of 1 Hz. From these data, for each player, we have extracted six characteristics capturing the minimum (hr_min), maximum (hr_max) and mean (hr_mean) value of heart rate, its standard deviation (hr_std), and the fraction of time each player kept their heart rate around their minimum value (hr_bottom) and around the maximum value (hr_peak).

The gameplay events extracted contain information about the duration of each run, the number of stages cleared, and the number of defeats in the game. The features extracted from the gameplay data are the time length of each run (*mean_run_time*), the number of stages visited in total (*total_chambers*), the furthest stage reached in every run (*max_chamber*), the number of defeats and runs (*number_of_runs*) and the number of times a player reaches the first boss encounter (*boss_encounters*).

4.2 Analysis

As a first approach to answer the research questions presented in Sect. 1, we conducted a correlation analysis using Pearson's correlation (r) to identify the presence of a linear relationship between the scales measuring the frustration tolerance, the heart rate features, the in-game behaviour and the reported player experience.

Figure 2 shows the results of this analysis. The correlations show how each group of features and scales is internally correlated, with player experience aspects such as *Flow* being strongly correlated with *Positive experience* ($r > 0.7$).

Across the scales, we can observe how the behavioural feature *max_chamber* shows a medium-strong correlation ($r > 0.6$) with multiple scales in GEQ (*Competence* and *Positive Experience*) which is likely due to the tight relationship between enjoyment and performance in an action game like Hades.

It is interesting to observe how performance, described by *max_chamber*, demonstrates a medium correlation with both player experience scales and heart rate (*hr_std* and *hr_bottom*). This relationship emerges also by dividing the players according to their performance. When the players are split according to the median value of *max_chamber* (*median* = 9), the players with a higher performance than the threshold show a significantly higher *hr_std* (7.4 vs 4.6) and a significantly lower *hr_bottom* (16% vs 36%) $p < .001$.

Heart rate variability is an established measure of the activation of the sympathetic and parasympathetic nervous systems [17], so a likely interpretation is that the capacity to increase and decrease the heart rate is a sign of the player's ability to respond to the challenge and perform better in the game.

The scores resulting from the frustration tolerance scales do not show any strong correlation with the other scale types and features; however, there is a medium correlation between *Reward Responsiveness* and *Reported Challenge* that could indicate the role of the individual importance of reward with their perception of challenge.

Another possible impact of frustration tolerance and heart rate on player experience is their potential role as moderators on other features, i.e. whether combining individual differences with gameplay features yields a stronger correlation with player experience. To evaluate this kind of effect we trained a linear regression model using either heart rate or frustration tolerance features paired with gameplay features to assess the correlation of the linear combinations with reported player experience.

This analysis reveals that *hr_bottom* linearly combined with player performance (*max_chamber*) is strongly and significantly correlated with *Competence* ($r > 0.7$, $p < .001$) with a 16% improvement over the single factor correlation shown in Fig. 2. This result indicates that the differences in the dynamic behaviour of the heart rate play a moderator role in the players' feeling of competence.

In contrast, the same analysis performed on frustration tolerance scales does not reveal a strong combined correlation.

5   Discussion and Conclusions

In this paper, we present a study investigating the role of individual differences in the quantification of player experience. The individual differences accounted for in the study include players' heart rate behaviour and their self-reported response to frustration and stress, and the player experience is quantified through GEQ and the players' in-game behaviour.

The results and the analysis show no conclusive evidence of either an indirect or a direct impact of the reported frustration tolerance on the players' behaviour and player experience; however, the few reported medium correlations and moderator effects suggest this aspect might require some further investigation. It is important here to mention that the GEQ has received criticism that the factor structure is not stable [11]. In particular, the core factor *negative affect* was problematic, as it was not clearly separated from similar factors *challenge* and *tension*. This is likely part of the explanation for why the results here are not clearer. We encourage further exploration of this topic in later studies. In particular, investigating how GEQ correlates with physiological measures and in-game behaviour in other settings than the one presented here would be an interesting undertaking.

Individual differences in heart rate behaviour, instead, show both a medium correlation with player performance and a strong correlation as a moderator effect between player performance and self-reported competence.

These results are consistent with previous studies on the relationship between the activation of the autonomous nervous system and heart rate variability [17] and hint at the potential for heart rate dynamics to be a valuable feature for player segmentation and interpretation of players' experience. However, heart rate standard deviation has limited granularity compared to heart rate variability (HRV) [17], so to confirm the results of this study and draw a more accurate picture, we believe that further studies are needed using a measure with higher resolution. Additionally, this study was conducted on a particular game, Hades, to generalise findings; studies should include a broader range of games within the action game genre.

References

1. Burelli, P., Triantafyllidis, G., Patras, I.: Non-invasive player experience estimation from body motion and game context. In: 2014 IEEE Conference on Computational Intelligence and Games, pp. 1–7. IEEE (2014)
2. Canossa, A., Drachen, A., Sørensen, J.R.M.: Arrrgghh!!!: blending quantitative and qualitative methods to detect player frustration. In: Proceedings of the 6th International Conference on Foundations of Digital Games, Bordeaux France, June 2011, pp. 61–68. ACM (2011)
3. Carver, C.S., White, T.L.: Behavioral inhibition, behavioral activation, and affective responses to impending reward and punishment: the BIS/BAS scales. J. Pers. Soc. Psychol. **67**(2), 319–333 (1994)
4. Chang, B., Chen, S.Y., Jhan, S.-N.: The influences of an interactive group-based videogame: cognitive styles vs. prior ability. Comput. Educ. **88**, 399–407 (2015)
5. Christopoulos, G.I., Uy, M.A., Yap, W.J.: The body and the brain: measuring skin conductance responses to understand the emotional experience. Organizational Res. Meth. **22**(1), 394–420 (2019)
6. Davidson, S.: A multi-dimensional model of enjoyment: development and validation of an enjoyment scale (enjoy). Doctoral Dissertations and Master's Theses, April 2018
7. Drachen, A., Nacke, L.E., Yannakakis, G., Pedersen, A.L.: Correlation between heart rate, electrodermal activity and player experience in first-person shooter games. In: Proceedings of the 5th ACM SIGGRAPH Symposium on Video Games, Los Angeles California, July 2010, pp. 49–54. ACM (2010)
8. Supergiant Games. Hades, September 2020
9. IJsselsteijn, W.A., de Kort, Y.A.W., Poels, K.: The Game Experience Questionnaire. Technische Universiteit Eindhoven, Eindhoven (2013)
10. Larche, C.J., Musielak, N., Dixon, M.J.: The candy crush sweet tooth: how 'near-misses' in candy crush increase frustration, and the urge to continue gameplay. J. Gambl. Stud. **33**(2), 599–615 (2017)
11. Law, E.L.-C., Brühlmann, F., Mekler, E.D.: Systematic review and validation of the game experience questionnaire (GEQ) - implications for citation and reporting practice. In: Proceedings of the 2018 Annual Symposium on Computer-Human Interaction in Play, Melbourne, VIC Australia, October 2018, pp. 257–270. ACM (2018)
12. Meindl, P.: A brief behavioral measure of frustration tolerance predicts academic achievement immediately and two years later. Emotion **19**(6), 1081–1092 (2019)
13. Perez Martínez, H., Garbarino, M., Yannakakis, G.N.: Generic physiological features as predictors of player experience. In: D'Mello, S., Graesser, A., Schuller, B., Martin, J.-C. (eds.) ACII 2011. LNCS, vol. 6974, pp. 267–276. Springer, Heidelberg (2011). https://doi.org/10.1007/978-3-642-24600-5_30
14. Phan, M.H., Keebler, J.R., Chaparro, B.S.: The development and validation of the game user experience satisfaction scale (GUESS). Hum. Fact. J. Hum. Fact. Erg. Soc. **58**(8), 1217–1247 (2016)
15. Seymour, K.E., Macatee, R., Chronis-Tuscano, A.: Frustration tolerance in youth with ADHD. J. Atten. Disord. **23**(11), 1229–1239 (2019)
16. Shaker, N., Asteriadis, S., Yannakakis, G.N., Karpouzis, K.: Fusing visual and behavioral cues for modeling user experience in games. IEEE Trans. Cybern. **43**(6), 1519–1531 (2013)

17. Sztajzel, J.: Heart rate variability: a noninvasive electrocardiographic method to measure the autonomic nervous system. Swiss Med. Wkly. **134**, 514–522 (2004)
18. Tondello, G.F., Nacke, L.E.: Player characteristics and video game preferences. In: Proceedings of the Annual Symposium on Computer-Human Interaction in Play, Barcelona Spain, October 2019, pp. 365–378. ACM (2019)
19. Vecchione, M., Ghezzi, V., Alessandri, G., Dentale, F., Corr, P.J.: BIS and BAS sensitivities at different levels of personality description: a latent-variable approach with self- and informant-ratings. J. Pers. Assess. **103**(2), 246–257 (2021)
20. Wiemeyer, J., Nacke, L., Moser, C., 'Floyd' Mueller, F.: Player experience. In: Dörner, R., Göbel, S., Effelsberg, W., Wiemeyer, J. (eds.) Serious Games, pp. 243–271. Springer, Cham (2016). https://doi.org/10.1007/978-3-319-40612-1_9
21. Wright, K.A., Lam, D.H., Brown, R.G.: Reduced approach motivation following nonreward: extension of the BIS/BAS scales. Pers. Individ. Differ. **47**(7), 753–757 (2009)
22. Yannakakis, G.N., Martinez, H.P., Garbarino, M.: Psychophysiology in games. In: Karpouzis, K., Yannakakis, G.N. (eds.) Emotion in Games. SC, vol. 4, pp. 119–137. Springer, Cham (2016). https://doi.org/10.1007/978-3-319-41316-7_7
23. Yannakakis, G.N., Spronck, P., Loiacono, D., André, E.: Player modeling. In: Lucas, S.M., Mateas, M., Preuss, M., Spronck, P., Togelius, J. (eds.) Artificial and Computational Intelligence in Games, volume 6 of Dagstuhl Follow-Ups, pp. 45–59. Schloss Dagstuhl-Leibniz-Zentrum fuer Informatik, Dagstuhl, Germany (2013). ISSN 1868-8977
24. Yun, C., Shastri, D., Pavlidis, I., Deng. Z.: O' game, can you feel my frustration?: improving user's gaming experience via stresscam. In: Proceedings of the SIGCHI Conference on Human Factors in Computing Systems, Boston MA USA, April 2009, pp. 2195–2204. ACM (2009)
25. Zohaib, M.: Dynamic Difficulty Adjustment (DDA) in computer games: a review. Adv. Hum. Comput. Interact. **2018**, 1–12 (2018)

Math, Games, and Puzzles

Analysis of a Collatz Game and Other Variants of the $3n + 1$ Problem

Ingo Althöfer[1]($\boxtimes$), Michael Hartisch[2], and Thomas Zipproth[1,2]

[1] Friedrich Schiller University Jena, Jena, Germany
ingo.althoefer@uni-jena.de
[2] University of Siegen, Siegen, Germany
michael.hartisch@uni-siegen.de

Abstract. Introduced and discussed are new variants of the $3n + 1$ problem, in particular a 2-player game with moves $3n + 1$ and $3n - 1$ and subsequent halving, where the player is winner who first reaches the 1. Variants of the classical $3n + 1$ problem are discussed with respect to a general convergence conjecture. Finally, we set up prizes for solutions of the Collatz problem and simpler variants.

Keywords: Collatz conjecture · Collatz game · $3n + 1$ problem · $3x + 1$

1 Introduction

In 1937 Lothar Collatz invented the $3n + 1$ problem [Ogi72, Gar72]: Starting with an initial odd natural number n, we calculate $3n + 1$ and proceed with halving steps until we encounter another odd number. This iterative process is then repeated. Can we be certain that this sequence of numbers eventually converges to 1? The special role of the number 1 becomes obvious by looking at the resulting sequence when starting from there:

$$1 - (4 - 2-)1$$

Hence, 1 is a fixed point of this procedure, i.e. as soon as 1 is reached the sequence degenerates to an infinite sequence of ones. Note that in our representation, we omit the even numbers that are encountered in the sequence, but we include them in these initial examples within parentheses for illustration. A less trivial example starts at 7:

$$7 - (22-)11 - (34-)17 - (52 - 26-)13 - (40 - 20 - 10-)5 - (16 - 8 - 4 - 2-)1$$

More formally we can define the halving function as

$$h : \mathbb{N} \to \mathbb{N}, \ h(x) = \frac{x}{2^{p(2,x)}},$$

where $p(2, x)$ is the number of occurrences of the factor 2 in the prime factorization of x. Hence, the sequence of (odd) numbers arising when starting from the odd number n_0 is n_0, $n_1 = h(3n_0 + 1)$, ..., $n_{t+1} = h(3n_t + 1)$, ...

© The Author(s), under exclusive license to Springer Nature Switzerland AG 2024
M. Hartisch et al. (Eds.): ACG 2023, LNCS 14528, pp. 123–132, 2024.
https://doi.org/10.1007/978-3-031-54968-7_11

Conjecture 1 (Collatz conjecture). For any odd starting value $n_0 \in \mathbb{N}$ the sequence $(h(3n_t + 1))_{t \in \mathbb{N}_0}$ leads to 1 in finitely many steps.

So far, no proof or counterproof has been found. But computational experiments have shown for all starting values until $1.77 \cdot 10^{21}$ that Conjecture 1 holds[1]. Most mathematicians believe the problem to be very difficult. For instance, Paul Erdős said in the 1980's: "Hopeless, absolutely hopeless. Mathematics is not ripe yet for such problems" [Guy04]. In the extensive book by Jeff Lagarias further theory, comments and results regarding the Collatz problem can be found [Lag10]. Rather recently, Terence Tao has achieved the best partial result so far [Tao19].

We do not tackle the $3n + 1$ problem directly, but propose a new 2-player game related to the Collatz problem. The players move in turn. Starting point is an odd natural number $n_0 > 1$. A move means to build either $3n + 1$ or $3n - 1$, and then to half, until an odd number is reached, i.e. in each move the player in turn has only two possible choices and the next player's move starts at $n_{t+1} \in \{h(3n_t + 1), h(3n_t - 1)\}$. The player who first reaches 1 at the end of his or her move wins the game, or in other words: the player starting from $n_t = 1$ for the first time loses. As an example assume a starting value 3. The two possible moves are $h(3 \cdot 3 + 1)$ and $h(3 \cdot 3 - 1)$, resulting in 5 and 1, respectively. Hence, by selecting the move $3 \cdot 3 - 1$ the player in turn wins immediately. Similarly, starting from 5 the player in turn can win by playing $h(3 \cdot 5 + 1) = h(16) = 1$. Starting at 7, however, results in a loss if the opponent reacts perfectly: The two options are $h(3 \cdot 7 + 1) = h(22) = 11$ and $h(3 \cdot 7 - 1) = h(20) = 5$. As shown above the position with starting value 5 is a winning position for the opponent and the same holds for 11 when choosing $h(3 \cdot 11 - 1) = 1$.

In Sect. 2 we show the computational results for the game with moves $3n - 1$ and $3n + 1$, and briefly discuss two game variants in Sect. 3. In Sect. 4 variants of the Collatz problem are discussed and Sect. 5 deals with a Markov model related to the Collatz problem. We conclude in Sect. 6 by listing open problems and advertising monetary prizes.

2   The $3n \pm 1$ Game

The $3n \pm 1$ game is a two-player zero-sum game, where the state of the game is given by the player in turn and the current value n_t. A player has two move options: Computing $3n_t + 1$ or $3n_t - 1$ and subsequently halving the resulting number, until the odd number n_{t+1} emerges, i.e. the options are leaving the other player with value $n_{t+1} = h(3n_t + 1)$ or $n_{t+1} = h(3n_t - 1)$. What is special about this game is that the value 1 might never be reached. For instance, if both players want to go on forever, they may realize the following pattern: Only one of the two numbers $3n - 1$ and $3n + 1$ can be divided by 4 or more. Playing the other option means going from n_t to $n_{t+1} > n_t$. Nevertheless, standard retrograde analysis can be used to examine optimal play [Str70, Tho86]. We implemented

---

[1] See http://www.ericr.nl/wondrous/index.html for further interesting facts.

an iterative procedure that updates values on game positions until it converges. When trying to compute the optimal game play up to starting number n_{max} we in fact only want to compute about $\frac{1}{2} n_{max}$ values as only odd numbers can be starting numbers. During the process, care has to be taken to catch those game paths which lead to positions above n_{max}. For this reason, we have to extend the state space to incorporate much larger starting numbers up to $M \cdot n_{max} \gg n_{max}$ which have to be investigated.

In our computational analysis we selected $n_{max} = 999{,}999$ and $M = 150$. In preliminary tests it had turned out that $M = 100$ was not sufficient to determine optimal results for all positions below 1 million. Our analysis has shown that all games with odd starting values smaller than one million terminate at 1 earlier or later if both players play optimally. Hence, we state the following conjecture.

Conjecture 2. In case of optimal play by both players in the $3n \pm 1$ game all starting values lead to 1.

Furthermore, we observed that 53% of the starting values constitute winning positions. Of course, perfect play shall also mean that the winner tries to win as quickly as possible, and the loser tries to postpone the loss as long as possible. This rule was applied when selecting the optimal successor. In Fig. 1 we show statistics regarding the game lengths. Note, however, that the computing process was terminated as soon as for every position up to n_{max} the outcome (win/loss) was clear. Therefore, the number of moves especially for large starting values might not yet be optimal: for winning positions only one of the successors needs to result in a win, while the other move might also result in a (not yet detected win) with fewer moves.

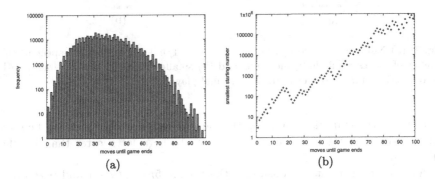

Fig. 1. (a) Number of starting values of the $3n \pm 1$ game with respect to the number of moves until the games ends and (b) smallest starting number found with the respective number of moves during optimal play. Note the logarithmic y-axis in both figures.

3   Game Variants

There is an unlimited number of possible variations of this game when generalizing the two possible moves to $an+b$ and $xn+y$ for odd numbers $a, b, x, y \in \mathbb{Z}$. For two variants we briefly showcase computational insights and establish conjectures.

First we consider the game with moves $3n+1$ and $n+1$. Here, starting values 3, 5 and 7 are winning positions, while starting value 9 (leading to 7 and 5 in the two possible moves) is a loss for the player in turn. Again, we perform the same analysis as before. And analogously to the $3n \pm 1$ game we were able to compute the optimal outcomes if both players play optimally for all starting values up to $n_{\max} = 10^6$. To that end we set $M = 100$, which had not been enough in the basic version of the game. It turns out that about 62% of starting values are winning positions.

Conjecture 3. In case of optimal play by both players in the $3n + 1/n + 1$ game all starting values lead to 1.

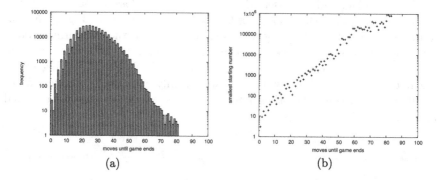

(a)   (b)

Fig. 2. (a) Number of starting values of the $3n + 1/n + 1$ game with respect to the number of moves until the games ends and (b) smallest starting number found with the respective number of moves during optimal play.

In Fig. 2 the statistics regarding the game length are shown. It can be seen that for this variant, games on average tend to be a lot shorter than for the $3n \pm 1$ game.

As a second variant we considered the version $5n + 1/n + 1$ and performed the same computational analysis. However, even for $n_{\max} = 1,000$ and $M = 10,000$ we observed, that no general convergence occurred: while starting value 3 constitutes a winning position, already no optimal move for starting value 5 was computed. The reason for this seems to be that if both players always select move $h(5n + 1)$ the resulting sequence of values tends to be increasing, which cannot be evaluated in general using our approach. In particular, if the other option of playing $h(n + 1)$ also is not a winning move (or also unknown) the option of just playing a move resulting in potential infinite play is preferred. Hence, we formulate the following conjecture.

Conjecture 4. In the $5n + 1/n + 1$ game almost all starting values will *not* lead to an end in value 1, if both players play optimally.

4   Collatz Variants and a General Conjecture on Convergence

We now consider a more general Collatz conjecture based on s odd numbers $k_0, \ldots, k_{s-1}$. We start with an odd number $n_0 \in \mathbb{N}$ which is mapped to $n_1 = h(k_0 \cdot n_0 + 1)$. We construct a sequence of numbers $(n_0, n_1, \ldots)$ where for any stage $t \geq 0$ the successor n_{t+1} is computed via

$$n_{t+1} = h(k_{\mathrm{mod}(t,s)} n_t + 1) ,$$

where $\mathrm{mod}(t, s)$ is the remainder of the division $\frac{t}{s}$, i.e. the odd numbers $k_0, \ldots, k_{s-1}$ are traversed in the given order repeatedly to serve as factors for the multiplication. We call the completion of s such function calls a *round*. Note that this general setting also contains the classic Collatz problem with $s = 1$ and $k_0 = 3$. Furthermore, observe that for instance $k_0, k_1, k_2, \ldots, k_{s-1}$ may give another variant than $k_1, k_0, k_2, \ldots, k_{s-1}$, if $k_0 \neq k_1$.

We define $C = k_0 \cdot k_1 \cdots k_{s-1}$ and we call $D = \frac{C}{4^s}$ the decisive multiplier. Note that $D \neq 1$ for any choice of $k_0, \ldots, k_{s-1}$, because C is an odd number.

Conjecture 5. Given any sequence of odd numbers $k_0, \ldots, k_{s-1}$. If $D < 1$, then each sequence starting from any odd starting value n_0 runs into one of finitely many cycles, where the set of cycles depends on the choice of $k_0, \ldots, k_{s-1}$.

Conjecture 6. If $D > 1$, then almost all starting values run to infinity, i.e. the asymptotic density of starting values resulting in infinitely growing sequences is 1.

A simple intuitive argument why this conjecture is reasonable is that the halving function on average performs two division by 2, assuming a uniform distribution of the even numbers put into the halving function. Thus, stage t contributes an approximate average factor of $\frac{k_t}{4}$, which means an approximate growth rate of D for every round.

We performed several computational experiments for various cases. We first demonstrate the case with $s = 2$ and factors $k_0 = 3$ and $k_1 = 5$. The decisive value is $D = \frac{15}{16} < 1$ and hence we expect convergence to a cycle for every starting number. Computational runs for all starting values up to 361 with 128-bit arithmetic gave the results found in Table 1. In all cases the sequence ended in one of the three cycles $(7, 11)$, $(9, 7)$, or $(13, 5)$. In each row, first the starting value is given, then the number of rounds, and the largest occurring value during the computation. Here it was most impressive to observe an intermediate value of nearly $3 \cdot 10^{12}$ for $n_0 = 95$.

Table 1. Selected results for $s = 2$, $k_0 = 3$, and $k_1 = 5$.

n_0	rounds	$\max_{t \in \mathbb{N}_0} \left(k_{\mathrm{mod}(t,s)} n_t + 1 \right)$
5	1	16
7	1	56
9	1	28
11	5	1,840
13	1	40
15	3	116
...	...	...
91	20	1,226,576
93	6	1,840
95	321	2,958,675,001,976
97	6	1,840
...	...	...
355	30	986,976
357	2	1,072
359	20	19,176
361	4	4,776

As a second setting we considered $s = 2$ with $k_0 = 15$ and $k_1 = 1$, again with $D = 15/16 < 1$. Using 128-bit arithmetic we checked all odd starting values until $1,797$ (and got an overflow for $1,799$). In all cases convergence to some finite cycle occurred. Selected data on interesting starting values can be found in Table 2.

Table 2. Selected results for $s = 2$, $k_0 = 15$, and $k_1 = 1$.

n_0	rounds	$\max_{t \in \mathbb{N}_0} \left(k_{\mathrm{mod}(t,s)} n_t + 1 \right)$
1	1	16
7	4	766
15	5	856
23	16	800,056
29	173	643,680,766
133	175	643,680,766
1787	266	42,897,306,732,256

We also performed experiments for the setting $s = 3$ and odd numbers $k_0 = 3$, $k_1 = 3$, and $k_2 = 7$ which gives $D = \frac{63}{64} < 1$. With 256-bit-arithmetic we were able to compute results for starting values until $n_0 = 2,797$, for which the most impressive values are given in Table 3. An overflow occurred at $n_0 = 2,799$. However, with unlimited bit-arithmetic, the sequence starting from $n_0 = 2,799$ reached the value of 1 after $14,966$ rounds with highest value larger than $4.5 \cdot 10^{126}$. Using this unlimited bit-arithmetic even more impressive record holders where found, e.g. the sequence starting from $n_0 = 14,975$ reached the cycle $(93, 35, 53)$ in $51,639$ rounds where the largest number of the sequence is about $9.9 \cdot 10^{275}$.

Table 3. Selected results for $s = 3$, $k_0 = 3$, $k_1 = 3$, and $k_2 = 7$.

n_0	rounds	$\max_{t \in \mathbb{N}_0} \left(k_{\mathrm{mod}(t,s)} n_t + 1 \right)$
1	2	8
5	2	16
7	8	7,680
199	219	$>3.0 \cdot 10^{13}$
479	1,507	$>5.6 \cdot 10^{39}$
1,655	5,688	$>5.3 \cdot 10^{68}$
1,967	5,159	$>1.5 \cdot 10^{68}$

Some additional interesting results for other settings: For $s = 4$ with $k_0 = 1$, $k_1 = 3$, $k_2 = 5$, and $k_3 = 17$ we have $D = \frac{255}{256}$. For starting value $n_0 = 4,165$ it took $180,902$ rounds to reach the cycle $(747, 187, 281, 703)$ and there were numbers in the sequence larger than $2.1 \cdot 10^{387}$.

For the same setting the sequence starting with $n_0 = 6,021$ took $172,580$ rounds to reach the cycle $(2259, 565, 53, 133, 1131, 283, 425, 1063)$, and highest intermediate value of about $5.0 \cdot 10^{546}$.

All computations in this section were performed on a Haswell Core i7-4790K @ 4.00 GHz. For each setting computations were executed within the range of minutes.

5 Convergence in a Stochastic Variant of the Collatz Problem

Many people believe that the Collatz conjecture is true. This is based on the observation that for a randomly chosen odd number n the even number $3n + 1$ is divisible either by 2 (and not by 4) or by 4 (and not by 8) or by 8 or by 16 and so on. In particular, in any sequence of four consecutive even numbers two are divisible by 2 and not by 4, one is divisible by 4 and not by 8, and the fourth one is divisible by (at least) 8. On average, $3n + 1$ contains prime factor 2 with multiplicity 2, because the infinite series $\frac{1}{2} \cdot 1 + \frac{1}{4} \cdot 2 + \frac{1}{8} \cdot 3 + \ldots$ converges to 2. So, if numbers were "really random", each round using the $3n + 1$ rule implies multiplying by 3 and dividing by 4 (of course only approximately, due to the

'+1' term). But in reality it is not given for free that computing $h(3n+1)$ yields sequences of "uniformly distributed random odd numbers".

We want to shine light on this by analysing a stochastic variant of the Collatz problem. Namely, one of the following random actions is taken on the odd number n: build $3n-1$, $3n+1$, $3n+3$, $3n+5$, each with probability $\frac{1}{4}$, independently of the history. Then do down-halving, resulting in a new odd number.

$3n-1$, $3n+1$, $3n+3$, $3n+5$ is a quadruple of successive even numbers as described above. Choosing each one of them with probability $\frac{1}{4}$ means that on average at least $\frac{1}{2} \cdot 1 + \frac{1}{4} \cdot 2 + \frac{1}{4} \cdot 3$ halving steps will happen. For asymptotic convergence it is important that the average, i.e. the geometric mean G, has to be smaller than 1. This is the case, as $G \leq \frac{3}{2} \cdot \frac{3}{2} \cdot \frac{3}{4} \cdot \frac{3}{8} = \frac{81}{128} < 1$.

Applying the theory of random walks with drift [Woe00] it is easy to see that large starting values will become small with probability 1 in the long run. Appendix A contains a discussion of a basic model.

The convergence argument from above does not give a proof for the simpler stochastic model where n becomes $3n-1$ and $3n+1$ with probability $\frac{1}{2}$ each, as in this case we only know $G^\star \leq \frac{3}{2} * \frac{3}{4} = \frac{9}{8}$, which is unfortunately larger than 1. By the way, in the stochastic model we chose $3n-1$, $3n+1$, $3n+3$, and $3n+5$ instead of the more symmetric setting with $3n-3$, $3n-1$, $3n-1$, and $3n+3$, because $3n-3$ would leave the set of positive integers for $n=1$.

6   Discussion, Open Problems, Prizes

We discussed game versions that build upon the famous Collatz problem and we furthermore discussed variants of the $3n+1$ problem including stochastic versions. We stated several conjectures and collected some clues that might help tackling these conjectures.

There are several open problems beyond our stated conjectures.

- How could a proof of convergence look for a stochastic version of the Collatz problem, where in each step $3n+1$ and $3n-1$ are selected both with probability $\frac{1}{2}$?
- Regarding the general conjecture proposed in Sect. 4, further experiments with other settings will be interesting.
- How similar are these structures, if instead of $k_j \cdot n_t + 1$ for all $j \in \{0, \ldots, s-1\}$ some moves are of the type $k_j \cdot n_t - 1$ or $k_j \cdot n_t + c$ for some suitable odd constant c?
- What will happen in Collatz games, where the two players have different options? One example: Player A can move $3n+1$ and $n+1$, Player B can move $3n-1$ and $n-1$. Who of the two players has better chances, if the game starts with some very large random odd number?

Finally, we would like to draw attention to monetary prizes announced by Ingo Althöfer for the solution of the following problems[2]:

---

[2] Prizes offered only for solutions submitted until December 31, 2037. Prizes offered only for the first solution. Legal actions are excluded.

1. Prove or disprove that there exists some odd number $a \geq 5$ and an odd number n_0, such that starting value n_0 leads to infinity under the rule $n_{t+1} = h(a \cdot n_t + 1)$. (25 Euro)
2. Prove or disprove that the sequence starting from $n_0 = 1$ leads to infinity under the rule $n_{t+1} = h(9n_t + 1)$. (50 Euro)
3. Prove or disprove for the $3n \pm 1$ game that all starting values lead to 1 under optimal play by both players. (500 Euro)
4. Prove or disprove the original Collatz conjecture: for any odd $n_0 \in \mathbb{N}$ the sequence arising when using rule $n_{t+1} = h(3n_t + 1)$ leads to 1. (1000 Euro)

Further information, news and updates can be found online[3].

Acknowledgments. Thanks to Frank Brenner, Horst Wandersleben, Wolfgang Woess, and Eric Roosendaal for further computations and discussions. Furthermore, we thank the anonymous reviewers for their constructive comments.

A   Random Walks with Drift

Given is the discrete strip of natural numbers, where each number $i > 0$ has neighbors $i - 1$ and $i + 1$, and $i = 0$ is an absorbing state. Let there be a discrete set of times steps $t \in \mathbb{N}$. A particle starts at some number n and goes to a random neighbor in each discrete time step, independently of the former steps. In the case without drift the particle goes in time step t from $i > 0$ to $i - 1$ with probability 0.5 and to $i + 1$ also with probability 0.5, independently of its history. It is well known that the particle will reach state 0 sooner or later with probability 1, and that the expected number of steps until absorption is infinite, even for starting state $n = 1$ [Woe00].

In a model with uniform drift, there is some $\epsilon \in [-\frac{1}{2}, \frac{1}{2}]$, such that a particle standing on $i > 0$ goes to $i-1$ with probability $0.5 + \epsilon$ and to $i+1$ with probability $0.5 - \epsilon$. In case of an $\epsilon > 0$ the particle has a drift towards 0, and for $\epsilon < 0$ a drift towards infinity.

Theory says [Woe00]: For $\epsilon > 0$ and any starting number the particle will reach 0 with probability 1 earlier or later in finitely many steps. For $\epsilon < 0$ and starting number n_0 there exists a value $q(\epsilon, n_0) < 1$, such that the particle will be absorbed at 0 with probability $q(\epsilon, n_0)$ and goes to infinity with the remaining probability $1 - q(\epsilon, n_0)$.

In a more general setting the particle can for instance jump from i to $i - 3$, $i - 2$, $i - 1$, i, $i + 1$, $i + 2$, $i + 3$, and $i + 4$ with probabilities $p(-3)$, $p(-2)$, $p(-1)$, $p(0)$, $p(1)$, $p(2)$, $p(3)$, and $p(4)$, respectively. If $3p(-3) + 2p(-2) + 1p(-1) > 1p(1) + 2p(2) + 3p(3) + 4p(4)$, the particle is drifting towards the left, i.e. towards zero. In this case it will be absorbed by 0 (or negative numbers) with probability 1 in the long run.

In the stochastic Collatz variant with logarithmic scale the particle is approximately jumping from its current state n_t to $n_t + \log(\frac{3}{2})$, $n_t + \log(\frac{3}{4})$, $n_t + \log(\frac{3}{8})$,

---

[3] https://www.althofer.de/collatz-prizes.html.

$n_t + \log(\frac{3}{16})$, ... with probabilities $\frac{1}{2}$, $\frac{1}{4}$, $\frac{1}{8}$, $\frac{1}{16}$, ..., respectively. Observe that $\log(\frac{3}{4})$, $\log(\frac{3}{8})$,... are all negative numbers. And $\frac{1}{2}\log(\frac{3}{2}) + \frac{1}{4}\log(\frac{3}{4}) + \frac{1}{4}\log(\frac{3}{8}) = \frac{1}{4}\log(\frac{3^4}{2^2 \cdot 4 \cdot 8}) = \frac{1}{4}\log(\frac{81}{128}) < 0$. The jumping widths are only approximately $\log(\frac{3}{2})$, $\log(\frac{3}{4})$, $\log(\frac{3}{8})$, ..., due to the '+1' term in $3n + 1$. However, for large values of n_t, this deviation is almost meaningless.

References

[Gar72] Gardner, M.: Miscellany of transcendental problems-simple to state but not at all easy to solve. Sci. Am. **226**(6), 114 (1972)

[Guy04] Guy, R.K.: Unsolved Problems in Number Theory, 3rd edn., pp. 330–336. Springer, New York (2004)

[Lag10] Lagarias, J.C.: The Ultimate Challenge: The 3x+1 Problem. Monograph Books. American Mathematical Society (2010)

[Ogi72] Ogilvy, C.S.: Tomorrow's Math: Unsolved Problems for the Amateur. Oxford University Press (1972)

[Str70] Ströhlein, T.: Untersuchungen über kombinatorische Spiele. Ph.D. thesis, Technische Hochschule München (1970)

[Tao19] Tao, T.: Almost all orbits of the collatz map attain almost bounded values. arXiv e-prints, pages arXiv-1909 (2019)

[Tho86] Thompson, K.: Retrograde analysis of certain endgames. J. Int. Comput. Games Assoc. **9**(3), 131–139 (1986)

[Woe00] Woess, W.: Random Walks on Infinite Graphs and Groups. 138, Cambridge University Press (2000)

Implicit QBF Encodings for Positional Games

Irfansha Shaik[1] , Valentin Mayer-Eichberger[2], Jaco van de Pol[1(✉)] ,
and Abdallah Saffidine[3]

[1] Aarhus University, Aarhus, Denmark
{irfansha.shaik,jaco}@cs.au.dk
[2] University of Potsdam, Potsdam, Germany
valentin@mayer-eichberger.de
[3] The University of New South Wales, Sydney, Australia
abdallahs@cse.unsw.edu.au

Abstract. We address two bottlenecks for concise QBF encodings of maker-breaker positional games, like Hex and Tic-Tac-Toe. We improve a baseline QBF encoding by representing winning configurations implicitly. The second improvement replaces variables for explicit board positions by a universally quantified symbolic board position. The paper evaluates the size of these lifted encodings, depending on board size and game depth. It reports the performance of QBF solvers on these encodings. We study scalability up to 19×19 boards, played in human Hex tournaments.

1 Introduction

This paper presents new concise encodings of positional games in Quantified Boolean Logic (QBF). In these games, two players claim empty board positions in turns. Examples include Hex, Harary's Tic-Tac-Toe (HTTT), and Gomoku.

Quantifier alternations in QBF can naturally express the existence of a winning strategy of bounded depth. This allows solving these PSPACE-complete games using the sophisticated search techniques of generic QBF solvers. The quality (size and structure) of the encoding has a great influence on performance, but precise knowledge of what constitutes a good encoding is quite limited.

We address two bottlenecks in QBF encodings for positional games [23]: First, in some positional games, like Hex, winning configurations are defined in terms of paths of connected positions. So an explicit representation of the goal is exponential in the board size. We study implicit-goal representations, based on neighbor relations. implicit-goal constraints not only yield concise encodings but also appear to boost the performance of QBF solvers by an order of magnitude.

The other bottleneck is duplication of variables and clauses by instantiation to board positions and unrolling to bounded depth. Extending techniques from planning [30], we represent symbolic positions by universal variables. This lifted encoding is concise, at the expense of an extra quantifier alternation.

We report on 8 QBF encodings, combining all ideas mentioned above:

© The Author(s), under exclusive license to Springer Nature Switzerland AG 2024
M. Hartisch et al. (Eds.): ACG 2023, LNCS 14528, pp. 133–145, 2024.
https://doi.org/10.1007/978-3-031-54968-7_12

Goal: / Board:	Explicit	Lifted	Stateless
All minimal paths	3.1 (EA)	[29] (LA)	[29] (SA)
Neighbor-based	3.2 (EN)	4.1 (LN)	4.3 (SN)
Transversal game	3.2 (ET)	4.2 (LT)	–

We study the size of all encodings, depending on board size and game depth. We also measure the performance of existing QBF solvers on all encodings, applied to a benchmark of Piet Hein's Hex puzzles (3×3 - 7×7 boards) and to human-played Hex championship plays (19×19 board).[1]

2 Preliminaries

Maker-Breaker Positional Games and QBF. A positional game is played on a board, on which players Black and White occupy empty positions in turns. In the maker-breaker variant, a game is won by the first player (Black) if and only if the final set of black positions contains some winning set; otherwise, it is won by the second player (White). We define these games formally as follows.

Definition 1. *Given a set of positions $\mathcal{P}$, let $\eta = |\mathcal{P}|$. A maker-breaker positional game Π is a tuple $\langle I, \mathcal{W} \rangle$, with initial state $I = (I_b, I_w)$ s.t. $I_b, I_w \subseteq \mathcal{P}$ and $I_b \cap I_w = \emptyset$, and goal condition $\mathcal{W} \subseteq 2^{\mathcal{P}}$, consisting of the winning sets.*

We assume, without loss of generality, that $I_b = I_w = \emptyset$, possibly after some preprocessing (cf. Sect. 3.2). A single play is a sequence of moves (occupying positions) chosen by players in turns. We assume Black plays first and in a maximal play, the game ends with Black's turn.

Definition 2. *Given Π, a single play ϕ is a sequence of k positions $\langle \phi_1, \ldots, \phi_k \rangle$ chosen by each player alternatively. The black moves $\phi_b = \{\phi_i \mid 1 \le i \le k, i \text{ odd}\}$ and the white moves $\phi_w = \{\phi_i \mid 1 \le i \le k, i \text{ even}\}$. A play is valid when $\{\phi_b \cup \phi_w\} \cap \{I_b \cup I_w\} = \emptyset$ and $\phi_i \ne \phi_j$ for all $i \ne j$.*

Definition 3. *Given Π and a valid play ϕ, we say the play ϕ is won by Black if and only if there exists a set of positions $\text{win} \in \mathcal{W}$ such that $\text{win} \subseteq I_b \cup \phi_b$.*

The goal of all encodings in this paper is to decide if Black has a winning strategy, i.e., Black wins all valid plays, played according to this strategy. We only consider strategies for a bounded number of moves up to depth d.

Hex and Generalized Hex. Hex is a well-known positional game played on an $n \times n$-board of hexagons, such that each non-border position has six neighbors [13]. The game is won by Black if there is a black path connecting Black's two opposite borders. The *Hex Theorem* [10] states that on a completely

---

[1] Precise encoding and measurement details are available in a technical report [29].

filled board, Black has a winning connection if and only if White's borders are not connected by white stones.

Because of its simplicity and rich mathematical structure, Hex has provided a source of inspiration for the design and implementation of specialized solvers [2], as well as for theoretical considerations on computational complexity [4,5,27].

Definition 4. Generalized Hex *is a 2-player game between Short and Cut. An instance is a 3-tuple* $\langle G, s, e \rangle$, *with graph* $G = \langle \mathcal{P} \cup \{s, e\}, \mathcal{E} \rangle$, *and distinguished nodes* s *and* e. *The players take turns claiming nodes from* $\mathcal{P}$. *Short wins if he claimed all nodes on some path from* s *to* e. *Cut tries to prevent this.*

To simplify notation, let $\Gamma_s = \{v \mid (s, v) \in \mathcal{E}\}$ and $\Gamma_e = \{v \mid (v, e) \in \mathcal{E}\}$. Short's goal is to create a path from any node of Γ_s to any node of Γ_e.

Quantified Boolean Formulas. We consider closed QBF formulas in prenex normal form, i.e., of the form $Q_1 x_1 \cdots Q_n x_n(\Phi)$, where Φ is a propositional formula with variables in $\{x_1, \ldots, x_n\}$ and each $Q_i \in \{\forall, \exists\}$. Every such formula evaluates to True or False. QBF evaluation is a standard PSPACE-complete problem. The complexity increases with the number of quantifier alternations.

Several QBF solvers exist, which operate on QBF in either QDIMACS format (Φ is a set of clauses in conjunctive normal form), or in QCIR format [17] (Φ is a circuit with and/or-gates and negation). QDIMACS is more low-level and allows efficient operations, while QCIR preserves more structure, and is more readable. We use both QDIMACS, using Bule [18] for concise specifications, and QCIR, transforming it to QDIMACS using the standard Tseitin transformation [32], introducing one existential variable per gate. However, it has been reported [1] that this step could blur structural information that is beneficial for QBF solvers.

3   Maker-Breaker Explicit-Board Encodings

We encode that Black wins from the empty board in d moves starting and ending with Black (so d is odd). Using QBF, we can naturally capture the moves of the players in a maker-breaker positional game by d alternating existential and universal variables. We first improve the corrective encoding COR by [23] (which was provided in QDIMACS) and then introduce implicit-goal constraints.

The main idea is to unroll the transition relation. In COR, board positions were maintained after each move, to test the validity of the moves and the winning condition. We save many frame conditions by maintaining board positions after black moves only. All moves are encoded logarithmically (in COR only the White moves). We identify the set of positions $\mathcal{P}$ with the integers $\{0, 1, \ldots, \eta - 1\}$. For $v \in \mathcal{P}$, we write $\overline{v}$ (and $\underline{v}$) for the set of bits assigned 1 in v (resp. 0 in v). For instance, if $v = 5 = b00\ldots 101$, we have $\overline{v} = \{0, 2\}$ and $\underline{v} = \{1, 3, 4, \ldots, \lceil \lg \eta \rceil - 1\}$. We introduce alternating variables M^t for the chosen move at step t. P^t represents the board at odd steps t (after each black move). Here variables $\mathsf{M}^t = \{m_i^t \mid 0 \le i < \lceil \lg \eta \rceil\}$ and variables $\mathsf{P}^t = \{b_v^t, w_v^t \mid v \in \mathcal{P}\}$.

$$\exists \mathsf{M}^1 \mathsf{P}^1 \forall \mathsf{M}^2 \exists \mathsf{M}^3 \mathsf{P}^3 \ldots \forall \mathsf{M}^{t-1} \exists \mathsf{M}^t \mathsf{P}^t \ldots \exists \mathsf{M}^d \mathsf{P}^d \tag{1}$$

For each vertex $v \in \mathcal{P}$, odd time step t, and bit index j we get the following clauses, which specify the color of each position at odd time steps t, depending on the previous position and the last moves of both players. When the clauses refer to variables not in the prefix, e.g., b_v^{-1}, they are substituted with $\bot$.

$$w_v^{t-2} \rightarrow w_v^t \tag{2}$$

$$\bigwedge_{i \in \overline{v}} m_i^{t-1} \wedge \bigwedge_{i \in \underline{v}} \neg m_i^{t-1} \wedge \neg b_v^{t-2} \rightarrow w_v^t \tag{3}$$

$$w_v^t \rightarrow \neg b_v^t \tag{4}$$

$$m_j^t \wedge \neg b_v^{t-2} \rightarrow \neg b_v^t \qquad \text{if } j \in \underline{v} \tag{5}$$

$$\neg m_j^t \wedge \neg b_v^{t-2} \rightarrow \neg b_v^t \qquad \text{if } j \in \overline{v} \tag{6}$$

Clause (2) propagates white positions over time; (3) turns a position white if a corresponding valid move was played; (4) ensures white positions are not black; Clauses (5, 6) ensure positions only become black by playing the corresponding move. If $\neg b_v^t$ is not forced, b_v^t will hold by the existential quantification over P^t. For a more detailed description, we refer to the technical report [24, (v2)].

3.1   Explicit-Board, All-Goals: (EA)

Similar to COR, we encode goal conditions using explicit winning configurations. We introduce existential variables $\exists \{h_h \mid h \in \mathcal{W}\}$ in the innermost existential block, indicating that Black won by configuration $h \in \mathcal{W}$. The clauses indicate that some winning configuration is completely black after d moves:

$$\bigvee_{h \in \mathcal{W}} h_h \tag{7}$$

$$h_h \rightarrow b_v^d \qquad \text{for } h \in \mathcal{W}, \text{ and } v \in h \tag{8}$$

The size of the encoding grows linearly with the number of winning configurations. Instead of all winning configurations, which can blow-up even for small boards in games like Hex, we consider minimal winning configurations for our encoding. Consider the Hex instance in Fig. 1. The set $\{d_1, d_2, d_3, c_3, c_4\}$ would be a winning path for Black, but it is subsumed by the winning subset $\{d_1, d_2, d_3, c_4\}$. Although EA only considers minimal winning paths, it still grows exponentially with board size for Hex (with a smaller exponent than COR).

3.2   Implicit-Goal Representations

For Hex, the winning condition is based on paths, whose number grows exponentially in the board size. In this section, we consider board games as graph games and use the neighbor edges to represent path conditions compactly. This technique is applicable to any positional game, like HTTT ([6,8,29]).

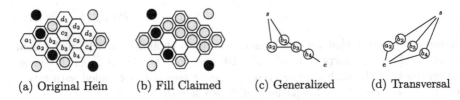

(a) Original Hein   (b) Fill Claimed   (c) Generalized   (d) Transversal

Fig. 1. Hein puzzle 9 and its reduction to Generalized Hex for $d = 5$. Black wins in (a) iff Black wins in (b) iff Short wins in (c) iff Cut wins in (d) (all in 5 steps). (Color figure online)

Transformations to Generalized Hex. Consider the Hex puzzle in Fig. 1a due to Hein [13]. Black has a winning strategy of depth 7 starting with c_3. For the sake of a running example, we will assume we search for a strategy of depth $d = 5$. Let us call B the existence of a 5-move win for Black on Fig. 1a.

Any winning strategy of d-moves involves at most $\ell + 1 = \frac{d+1}{2}$ black moves, so any winning configuration of size $> \ell + 1$ can be removed. Similarly, we remove any non-minimal winning configuration. This may result in positions not occurring in any winning configuration, which may be claimed by White (Breaker). Our query B is equivalent to Black having a 5-move win in Fig. 1b.

In any Maker-Breaker game, positions already claimed by either player can be preprocessed. Any winning configuration containing a white-claimed position is removed from $\mathcal{W}$ and black-claimed positions can be removed from any winning configuration. Following the contraction of [4, Thm 2] for Generalized Hex, we remove any Cut-claimed vertex and its incident edges and any Short-claimed vertex and turn its neighborhood into a clique. Applying this contraction to our running example gives Fig. 1c where Short has a 5-move win if and only if B.

The Hex Theorem allows solving Hex through the *transversal game*. To verify if Black has won after d moves, one can fill the remaining empty cells for White and ensure that even then White has no connecting paths. For this approach, we can similarly fill unnecessary cells for White (Fig. 1b) and apply the contraction process to Generalized Hex with the two players swapped as in Fig. 1d.

Explicit-Board Neighbor-Based (EN). We apply the symbolic neighbor-based goal encoding to the explicit-board constraints. Keeping the quantification and transition constraints (Eq. 1–6), we replace the explicit goal constraints (7–8) by the following constraints. Recall that $\ell + 1 = \frac{d+1}{2}$ is the maximum length of a winning path. We introduce Boolean variables: $\exists \{ \mathsf{p}_v^i \mid v \in \mathcal{P}, 0 \leq i \leq \ell \}$. Here p_v^i codes that position v is the i-th position in a winning witness path.

$$\mathsf{p}_v^i \rightarrow \mathsf{b}_v^d \qquad\qquad \text{for } v \in \mathcal{P}, 0 \leq i \leq \ell \qquad (9)$$

$$\bigvee_{v \in \Gamma_s} \mathsf{p}_v^0 \qquad\qquad (10)$$

$$\mathsf{p}_v^i \rightarrow \bigvee_{(v,w) \in \mathcal{E}} \mathsf{p}_w^{i+1} \qquad\qquad v \in \mathcal{P} \setminus \Gamma_e, 0 \leq i < \ell \qquad (11)$$

$$\neg p_v^\ell \hspace{6cm} v \in \mathcal{P} \setminus \Gamma_e \hspace{2cm} (12)$$

Clause (9) specifies that all positions on the path should be black on the final board d. The path starts at border Γ_s (10) and ends at border Γ_e (12). Clause (11) states that the path is connected by graph edges in $\mathcal{E}$ (extended with "stutter" steps to compensate for paths of different lengths).

Explicit-Board Transversal-Based Goals (ET). In this subsection, we specify the winning condition in the transversal game. This can be applied to any positional game, but it is effective for Hex because specifying that White has no connecting path is easier than specifying that Black has a connecting path.

We introduce existential variables $\exists \{r_v \mid v \in \mathcal{P}\}$, which hold for all positions that are connected to Γ_s through positions that are white or empty (i.e., $\neg b_v^d$). Then we simply check that no Γ_e position is connected. The following clauses for ET are combined with the quantification and transition constraints (Eq. 1–6).

$$\neg b_v^d \rightarrow r_v \hspace{4cm} \text{for } v \in \Gamma_s \hspace{2cm} (13)$$

$$r_v \wedge \neg b_w^d \rightarrow r_w \hspace{3cm} \text{for } (v,w) \in \mathcal{E} \hspace{2cm} (14)$$

$$\neg r_v \hspace{5cm} \text{for } v \in \Gamma_e \hspace{2cm} (15)$$

4   Implicit-Board Encodings

The Explicit-Board encodings (E*) duplicate variables and constraints for each position after a black move. Note that in positional games, the validity of all moves can be checked independently for each individual position. We use this structure to generate a lifted encoding, representing position constraints symbolically on a universal position variable. This follows the lifted encoding for planning [30]. We present our lifted encodings in QCIR format to preserve the structure of constraints. This opens up comparison with QBF solvers that operate on QCIR format directly, and nicely complements our handcrafted QDIMACS instances of the E* encodings (Sect. 3). We consider both a Neighbor-based implicit-goal encoding (Sect. 4.1) and a Transversal-based goal encoding (Sect. 4.2). Inspired by Causal Planning [20], the final encoding avoids states entirely (Sect. 4.3).[2]

4.1   Lifted Neighbor-Based Encoding (LN)

We introduce variables $M = \{M^1, \ldots, M^d\}$, where M^t is a sequence of $\lceil \lg(\eta) \rceil$ Boolean variables representing one move at time step t. Given a single play, we check if all moves are valid by using a single symbolic position, represented by P, a sequence of $\lceil \lg(\eta) \rceil$ universal variables. When expanded, the universal branches of P correspond to all board positions. We use two variables for each time step to represent the state of the symbolic position. Let $S^t = \{o^t, w^t\}$, representing if

---

[2] [29, Appendix B] also provides lifted and stateless encodings with implicit-board encodings and explicit-goal conditions, for completeness.

the symbolic state is *occupied* and/or *white* at time step t. We introduce witness variables $W = \{W^0, \ldots, W^\ell\}$, i.e., a sequence of positions that should form a winning configuration for Black. Here $\ell = \frac{d-1}{2}$, so $\ell + 1$ is the maximum size of a witness set in $\mathcal{W}$. Each W^i is a sequence of $\lceil \lg(\eta) \rceil$ Boolean variables. Finally, we introduce neighbor variables N, a sequence of $\lceil \lg(\eta) \rceil$ variables representing any neighbor of symbolic position P.

$$\exists M^1 \forall M^2 \ldots \exists M^d \quad \exists W^0, \ldots, W^\ell \quad \forall P \quad \exists S^1, \ldots, S^{d+1} \quad \exists N \quad (16)$$

The corresponding Lifted Encoding for finding a winning strategy for Black has the following shape:

$$\neg o^1 \wedge BR(M) \wedge SRC(W) \wedge TRG(W) \wedge G_{\mathcal{E}}(N, W, P) \wedge G_{sb}(S^{d+1}, W, P) \wedge \quad (17)$$

$$\bigwedge_{t=1,3,\ldots,d} T_b^t(S^t, S^{t+1}, P, M^t) \wedge \bigwedge_{t=2,4,\ldots,d-1} T_w^t(S^t, S^{t+1}, P, M^t) \quad (18)$$

We refer to [29, App. B.1] for the definition of the auxiliary formulas. Here we just provide an overview: The initial board is empty, hence $\neg o^1$. This forces *every* node to be unoccupied, since it is under the universal quantifier $\forall P$. The moves are restricted (BR) to legal moves, when η is not a power of 2.

For the Goal condition, we use the neighbor relation from the Generalized Hex input. First, we specify that the first and last witness positions lie on the proper borders, $SRC(W) \wedge TRG(W)$. The witness positions must be black (G_{sb}). We also specify that adjacent positions in the witness path are connected in the graph $(G_{\mathcal{E}})$. In particular, if a witness position matches P in the current branch, the next witness position matches its symbolic neighbor N.

Finally, we specify the transition constraints for Black (T_b^t) and White (T_w^t) separately. For Black, the position in the universal branch of the chosen move must be empty at time step t and black at $t + 1$. In all other branches, we just propagate the position. For White, the position is only updated if it is empty in the universal branch of the chosen move. Otherwise, the position is propagated.

4.2   Lifted Transversal-Based Encoding (LT)

Next, we provide the lifted variant of the transversal-based encoding ET. We use the same move, state and symbolic position variables and the same constraints as LN, except for the goal condition. We check that White can disconnect Black's borders in the traversal game. Similar to ET, we introduce reachability variables $\exists r_0, \ldots, r_{\eta-1}$ (one per position). The corresponding Lifted Encoding for finding a winning strategy has this shape:

$$\exists M^1 \forall M^2 \ldots \exists M^d \quad \exists r_0, \ldots, r_{\eta-1} \quad \forall P \quad \exists S^1, \ldots, S^{d+1}$$

$$\neg o^1 \wedge BR(M) \wedge \bigwedge_{t=1,3,\ldots,d} T_b^t(S^t, S^{t+1}, P, M^t) \wedge \bigwedge_{t=2,4,\ldots,d-1} T_w^t(S^t, S^{t+1}, P, M^t) \wedge$$

$$Start(P, w^{d+1}, o^{d+1}, r) \wedge Connected(P, w^{d+1}, o^{d+1}, r) \wedge End(r)$$

The Lifted variant of the goal constraints in ET (cf. Eq. 13–15) has three main constraints checked in the final state $d + 1$ (defined in detail in [29, App. B.2]): (1) Start: Non-black nodes on the start boarder Γ_s are reachable; (2) Connected: A non-black position is reachable if one of its neighbors is reachable; (3) End: Nodes on the end boarder Γ_e are not reachable.

4.3 Stateless Neighbor-Based (SN)

In positional games, Black's moves are valid if and only if they are different from all previous moves. Also, the winning condition can be expressed completely in terms of Black's moves. In the maker-breaker case, we don't even need to check the validity of White's moves. We just ignore White's moves that are played outside the board or on occupied positions. This is correct because in positional games White cannot win by giving up a move.

Table 1. Alternation depth and size of all presented encodings.

Enc	Alt	# Variables	# Clauses	Enc	Alt	# Variables	# Gates
EA	d	$d\eta + |\mathcal{W}|$	$\frac{1}{2}d\eta\lg\eta + d|\mathcal{W}|$	SN	d	$\frac{3}{2}d\lg\eta$	$d^2\lceil\lg\eta\rceil + \frac{1}{2}d\eta$
EN	d	$\frac{3}{2}d\eta$	$\frac{1}{2}d\eta\lg\eta$	LN	$d+1$	$\frac{3}{2}d\lg\eta$	$4d\lg\eta + 4\eta$
ET	d	$d\eta$	$\frac{1}{2}d\eta\lg\eta + |\mathcal{E}|$	LT	$d+1$	$d\lg\eta + \eta$	$2d\lg\eta + 6\eta$

The move and witness variables are the same as in Sect. 4.1. We can now drop all other variables for the SN encoding. The main constraints are: (1) Black cannot overwrite previous moves. (2) Witness positions must be played by Black. (3) Adjacent position pairs in W are neighbors in $\mathcal{E}$, and form a winning path between Black's borders. Here $\mathrm{bin}(W, v)$ encodes: "the binary value of W is v".

$$\exists\, M^1 \,\forall\, M^2 \ldots \exists\, M^d \qquad \exists\, W^0, \ldots, W^\ell$$

$$\mathrm{BR}(M) \wedge \Big(\bigwedge_{t=1,3,\ldots,d} \bigwedge_{i<t} M^t \neq M^i \Big) \wedge \Big(\bigwedge_{i=0}^{\ell} \bigvee_{t=1,3,\ldots,d} W^i = M^t \Big) \wedge$$

$$\mathrm{SRC}(W) \wedge \mathrm{TRG}(W) \wedge \bigwedge_{i=0}^{\ell-1} \bigwedge_{v=0}^{\eta-1} \Big(\mathrm{bin}(W^i, v) \implies \bigvee_{(v,w)\in\mathcal{E}} \mathrm{bin}(W^{i+1}, w) \Big)$$

5   Implementation and Evaluation

We provide two tools to generate explicit-goal encodings for all maker-breaker games and implicit-goal encodings for Hex and HTTT. The first tool,[3] extending the COR encoding [23], allows an implicit-goal specification and generates

---

[3] Available at https://github.com/vale1410/positional-games-qbf-encoding.

the EA, EN, and ET encodings in QDIMACS. The second tool[4] generates the implicit-board encodings SN, LN, LT – and LA and SA [29, App. B] – in QCIR.

5.1 Size of the QBF Encodings

Table 1 shows the alternation depth, number of variables, and number of clauses or gates for the explicit and implicit-board encodings of Hex,[5] depending on the number of positions (η), the depth of the game (d), the size of the winning set $\mathcal{W}$, and the number of edges $\mathcal{E}$. Obviously, the explicit-goal encoding (EA) depends on $|\mathcal{W}|$, which is unfeasible for 19×19 Hex boards. The implicit-board versions (L*) are more concise than the explicit-board versions (E*), at the expense of an extra quantifier alternation. The stateless encoding saves this quantifier, but its size grows quadratically in the depth of the game.

In Table 2 we measure the number of variables and clauses generated by our tools for increasingly deeper games on an empty 19×19 board. These can only be generated for the implicit-goal encodings (*N,*T).

Table 2. Size of encodings on 19×19 Hex board

Enc.	Generated QBF Encodings (variables/clauses)			
	$d=45$	$d=91$	$d=181$	$d=361$
EN	25k/122k	50k/246k	100k/488k	199k/972k
ET	17k/100k	34k/200k	67k/395k	134k/785k
LN	5k/22k	9k/33k	17k/54k	32k/97k
LT	5k/18k	8k/25k	12k/37k	22k/63k
SN	53k/261k	167k/720k	560k/2182k	2027k/7342k

We translate the QCIR benchmarks to QDIMACS for a fair comparison. For SN, the quadratic growth in the game depth is clearly visible.

To study the effect of the encoding on the performance of QBF solvers, we experiment with Piet Hein's benchmark of Hex puzzles on small boards. We also try "shallow" end games from human championships on a 19×19 board.

5.2 Solver Performance for Various Encodings

Experiment Design. We perform three experiments to evaluate the performance of QBF solvers on our encodings. The first one uses a benchmark of 30 instances of Hein's Hex puzzles of board sizes 3×3 to 5×5 from [23], which we call *Hein-base*. The second one uses an extended benchmark of 10 Hein's Hex puzzles with board sizes 6×6 and 7×7, which we call *Hein-hard*. The third one runs on games from the best human players in the recent 19×19 Hex championship.[6] These games are resigned quite early. This yields 15 False instances (for depths 11–17). Then we finish the game using the Hex-specific solver Wolve [2] and roll back some moves. This provides 8 True instances (for depths 9–17). After pruning useless positions (cf. Sect. 3.2), the number of open positions in the first and second set of benchmarks is at most 40. The 19×19 boards have 10-146 open positions. Many open positions become useless and can be pruned.

---

[4] Available at https://github.com/irfansha/Q-sage.
[5] We display an asymptotically equivalent function. [29, App. C] shows the exact size.
[6] From https://www.littlegolem.net. See [29, Appendix A] for the selection process.

We try QBF solvers CAQE [26], DepQBF [21], CQESTO [16] and Quabs [14]. We also use several QBF preprocessors, in particular Bloqqer [3], HQSpre [33], and QRATpre+ [22]. Two important techniques in QBF preprocessing are variable elimination and variable expansion [15]. For the first experiment, we run all combinations of QBF preprocessors and solvers on all encodings of the 30 Hein-base instances. For each encoding, we select the two best combinations. We apply the same combinations in the second and third experiment.

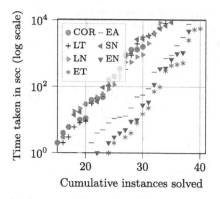

(a) Hein's instances solved by encodings.
3 hour time, 8GB mem limit (log scale).

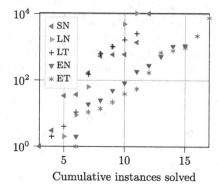

(b) Championship instances, lifted goals.
3 hour time, 32GB mem limit (log scale)

Fig. 2. Solving time for best solver-preprocessor combination per encoding

Experimental Results. We show cactus plots illustrating the cumulative number of solved instances over time, using the best solver/preprocessor for each encoding. Figure 2a shows the results for all Hein instances (base and hard) on all our six encodings and the original COR encoding. Figure 2b shows the results for the Championship benchmarks on the five implicit-goal descriptions.[7]
We now present our main observations: (1) The implicit-board encodings solve more instances than the original COR encoding, and the explicit-board encodings (E*) outperform COR significantly. For Hein's benchmarks, EA performs nearly as good as the implicit-goal versions EN and ET. However, for 19×19 boards we couldn't even generate the COR and EA encodings, since the set of winning paths cannot be practically enumerated. (2) Explicit-board encodings (E*) perform an order of magnitude faster than implicit ones (L*, S*). So the compact size of the latter doesn't compensate for the extra quantifier alternation. Still, LN solves the largest False instance uniquely. This championship instance has depth 15 and 83 open positions. (3) Within the implicit-board encodings, SN, LN and LT perform quite similarly, and none of them dominates the other. Among SN and LN, SN

---

[7] [29, Appendix D] gives detailed measurements per solver/preprocessor. [29, App. E] shows memory plots, and [29, Appendix F] separates True and False instances.

manages to solve a few more instances than LN overall. LT and LN complement each other in many cases, especially in the Championship benchmarks. (4) ET is the only encoding that solves all 40 instances of the Hein instances (up to depth 15 for False cases). It solves most instances of the second experiment: 11 False and 6 True instances. The unique True instance solved has depth 15 and 115 open positions, which is the largest solved instance overall.

6 Conclusion and Future Work

We addressed two bottlenecks for the QBF encoding of maker-breaker positional games. *Symbolic goal constraints* proved to be essential for generating QBF formulas for games with many winning configurations, such as Hex on a 19×19 board. On smaller boards, this technique also led to a boost in performance, leading to currently the fastest way to solve Hex puzzles using a generic solver. In fact, the implicit-goal representation can even be compressed logarithmically (at the expense of more quantifier alternations) by means of iterative squaring, as in bounded model checking and compact planning [7,19].

Using *symbolic board positions*, leads to even smaller, lifted QBF encodings, but current solvers cannot solve them as fast as the explicit-board representation. Still, many game instances could be solved with implicit-board representation. We submitted our encodings to the QBFeval 2022 competition [25].[8]

On a Hex 19×19 board, we could handle limited game depths, where many Hex board positions can be pruned away. We also investigated game positions where human champions resigned, because they believe the game is lost. Still, the actual number of moves required from these lost positions is quite high. Experts use several patterns specific to Hex (like ladders and bridges) to look ahead many steps. Even specialized Hex solvers [2] have difficulty solving these "obviously lost" positions, but the specialized solvers are of course much more efficient on Hex than generic QBF solvers on our encodings.

However, our approach is more general and can be applied to any maker-breaker positional game, as demonstrated by experiments for Harary's Tic-Tac-Toe in the full technical report [29, App. G] and thesis [28]. We refer to [24] for extensive experiments with the explicit-board encoding on other positional games. The ideas can easily be applied to maker-maker variants of positional games (for Hex, the two variants are equivalent). It is interesting for future work to apply similar techniques to non-positional games, like Connect-Four and Breakthrough (a first step is made in [31]). Another exciting direction is to investigate whether the concise modeling techniques for QBF in this paper extend to game approaches based on Quantified Integer Programs [9,11,12].

---

[8] Available at http://www.qbflib.org/qbfeval22.php.

References

1. Ansotegui, C., Gomes, C.P., Selman, B.: The Achilles' heel of QBF. In: AAAI 2005, pp. 275–281 (2005). http://dl.acm.org/citation.cfm?id=1619332.1619378
2. Arneson, B., Hayward, R., Henderson, P.: MoHex wins Hex tournament. ICGA J. **32**(2), 114 (2009)
3. Biere, A., Lonsing, F., Seidl, M.: Blocked clause elimination for QBF. In: Bjorner, N., Sofronie-Stokkermans, V. (eds.) CADE. LNCS, vol. 6803, pp. 101–115. Springer, Heidelberg (2011). https://doi.org/10.1007/978-3-642-22438-6_10
4. Bonnet, É., Gaspers, S., Lambilliotte, A., Rümmele, S., Saffidine, A.: The parameterized complexity of positional games. In: ICALP 2017, pp. 90:1–90:14 (2017)
5. Bonnet, É., Jamain, F., Saffidine, A.: On the complexity of connection games. Theor. Comput. Sci. (TCS) **644**, 2–28 (2016)
6. Boucher, S., Villemaire, R.: Quantified Boolean solving for achievement games. In: 44th German Conference on Artificial Intelligence (KI), pp. 30–43 (2021)
7. Cashmore, M., Fox, M., Giunchiglia, E.: Planning as quantified Boolean formula. In: ECAI 2012, pp. 217–222 (2012). https://doi.org/10.3233/978-1-61499-098-7-217
8. Diptarama, Yoshinaka, R., Shinohara, A.: QBF encoding of Generalized Tic-Tac-Toe. In: 4th IW on Quantified Boolean Formulas (QBF), pp. 14–26 (2016)
9. Ederer, T., Lorenz, U., Opfer, T., Wolf, J.: Modeling games with the help of quantified integer linear programs. In: van den Herik, H.J., Plaat, A. (eds.) ACG. LNCS, vol. 7168, pp. 270–281. Springer, Heidelberg (2011). https://doi.org/10.1007/978-3-642-31866-5_23
10. Gale, D.: The game of Hex and the Brouwer fixed-point theorem. Am. Math. Mon. **86**(10), 818–827 (1979)
11. Hartisch, M.: Quantified integer programming with polyhedral and decision-dependent uncertainty. Ph.D. thesis, Universität Siegen (2020)
12. Hartisch, M., Lorenz, U.: A novel application for game tree search - exploiting pruning mechanisms for quantified integer programs. In: Cazenave, T., van den Herik, J., Saffidine, A., Wu, I.C. (eds.) ACG. LNCS, vol. 12516, pp. 66–78. Springer, Heidelberg (2019). https://doi.org/10.1007/978-3-030-65883-0_6
13. Hayward, R.B., Toft, B.: Hex, the full story. AK Peters/CRC Press/Taylor (2019)
14. Hecking-Harbusch, J., Tentrup, L.: Solving QBF by abstraction. In: GandALF. EPTCS, vol. 277, pp. 88–102 (2018). https://doi.org/10.4204/EPTCS.277.7
15. Heule, M., Järvisalo, M., Lonsing, F., Seidl, M., Biere, A.: Clause elimination for SAT and QSAT. J. Artif. Intell. Res. (JAIR) **53**, 127–168 (2015)
16. Janota, M.: Circuit-based search space pruning in QBF. In: Beyersdorff, O., Wintersteiger, C. (eds.) SAT. LNCS, vol. 10929, pp. 187–198. Springer, Heidelberg (2018). https://doi.org/10.1007/978-3-319-94144-8_12
17. Jordan, C., Klieber, W., Seidl, M.: Non-CNF QBF solving with QCIR. In: AAAI-16 Workshop on Beyond NP (2016)
18. Jung, J.C., Mayer-Eichberger, V., Saffidine, A.: QBF programming with the modeling language Bule. In: Proceedings SAT 2022. Schloss Dagstuhl-Leibniz (2022)
19. Jussila, T., Biere, A.: Compressing BMC encodings with QBF. ENTCS **174**(3), 45–56 (2007). https://doi.org/10.1016/j.entcs.2006.12.022
20. Kautz, H.A., McAllester, D.A., Selman, B.: Encoding plans in propositional logic. In: Principles of Knowledge Representation and Reasoning (KR), pp. 374–384 (1996)

21. Lonsing, F., Egly, U.: DepQBF 6.0: a search-based QBF solver beyond traditional QCDCL. In: de Moura, L. (ed.) CADE. LNCS, vol. 10395, pp. 371–384. Springer, Heidelberg (2017). https://doi.org/10.1007/978-3-319-63046-5_23

22. Lonsing, F., Egly, U.: QRATPre+: effective QBF preprocessing via strong redundancy properties. In: Janota, M., Lynce, I. (eds.) SAT. LNCS, vol. 11628, pp. 203–210. Springer, Heidelberg (2019). https://doi.org/10.1007/978-3-030-24258-9_14

23. Mayer-Eichberger, V., Saffidine, A.: Positional games and QBF: the corrective encoding. In: Theory and Applications of Satisfiability Testing (SAT), pp. 447–463 (2020)

24. Mayer-Eichberger, V., Saffidine, A.: Positional games and QBF: a polished encoding. arXiv 2005.05098 (2023). https://doi.org/10.48550/arXiv.2005.05098

25. Pulina, L., Seidl, M., Shukla, A.: The 14th QBF solvers evaluation (QBFEVAL 2022) (2022). http://www.qbflib.org/QBFEVAL22_PRES.pdf

26. Rabe, M.N., Tentrup, L.: CAQE: a certifying QBF solver. In: Kaivola, R., Wahl, T. (eds.) Proceedings FMCAD 2015, pp. 136–143. IEEE (2015)

27. Reisch, S.: Hex ist PSPACE-vollständig. Acta Informatica **15**, 167–191 (1981)

28. Shaik, I.: Concise Encodings for Planning and 2-Player Games. Ph.D. thesis, Aarhus University (2023)

29. Shaik, I., Mayer-Eichberger, V., van de Pol, J., Saffidine, A.: Implicit state and goals in QBF encodings for positional games (extended version). arXiv 2301.07345 (2023). https://doi.org/10.48550/arXiv.2301.07345

30. Shaik, I., van de Pol, J.: Classical planning as QBF without grounding. In: ICAPS, pp. 329–337. AAAI Press (2022)

31. Shaik, I., van de Pol, J.: Concise QBF encodings for games on a grid. arXiv 2303.16949 (2023). https://doi.org/10.48550/ARXIV.2303.16949

32. Tseitin, G.S.: On the complexity of derivation in propositional calculus. In: Siekmann, J.H., Wrightson, G. (eds.) Automation of Reasoning, pp. 466–483. Springer, Heidelberg (1983). https://doi.org/10.1007/978-3-642-81955-1_28

33. Wimmer, R., Scholl, C., Becker, B.: The (D)QBF preprocessor HQSpre - underlying theory and its implementation. J. Satisf. Boolean Model. **11**(1), 3–52 (2019)

The Mathematical Game

Marc Pierre[(⊠)], Quentin Cohen-Solal, and Tristan Cazenave

LAMSADE, Université Paris Dauphine - PSL, CNRS, Paris, France
marcpierre1999@gmail.com

Abstract. Monte Carlo Tree Search can be used for automated theorem proving. Holophrasm is a neural theorem prover using MCTS combined with neural networks for the policy and the evaluation. In this paper we propose to improve the performance of the Holophrasm theorem prover using other game tree search algorithms.

1 Introduction

Monte Carlo Tree Search (MCTS) has been successfully applied to many games and problems [2]. It was used to build superhuman game playing programs such as AlphaGo [7], AlphaZero [8] and Katago [10]. It has been recently used to discover new fast matrix multiplication algorithms [3]. It is also used for automated theorem proving such as in the Holophrasm theorem prover [9]. In this paper, we propose to replace the MCTS used by Holophrasm by other game tree search algorithms. We will start by briefly explaining how metamath works, as well as the Holophrasm interface. Then we will move on to an application of existing tree search algorithms modified for this context, such as Minimax, PUCT or Product Propagation. Finally, we propose a new algorithm, which is an association of existing ones, and apply it in the context of Holophrasm.

2 Holophrasm and Metamath

2.1 Metamath

Metamath is a mathematics language [5]. Its main function is based on the principle of logical substitution. For example, let's imagine that we are at a certain step of a theorem's proof, and to move on to the next step we need to apply a proposition. To do this, we need to change the variables of the proposition we wish to apply, so that its hypotheses correspond to the current state of the proof. For more details, one can check the metamath's book ([5]). Holophrasm transforms this structure of theorem's proof into an "AND/OR" tree. OR nodes represents the current state of the proof. They are considered proven if one of their child is proven or if they are one of the initial hypotheses of the theorem. Their children are AND nodes representing a proposition to be applied to prove the OR father node. The children of an AND node are the set of hypotheses to be proven for the proposition to be true. A hypothesis is modeled by an OR

© The Author(s), under exclusive license to Springer Nature Switzerland AG 2024
M. Hartisch et al. (Eds.): ACG 2023, LNCS 14528, pp. 146–157, 2024.
https://doi.org/10.1007/978-3-031-54968-7_13

node and an AND node is considered proven if all its children are proven. In Holophrasm, the theorem is proved by working backward. The root being the conclusion of the proof modeled by an OR node, each of its children is an AND node which is a proposition with its substitution of variables. Concerning the theorems on which we will test our algorithms, the benchmarks are made up of a list of theorems and are provided by the Holophrasm interface. However, due to time constraints, we will only test on the first 200 theorems in the list.

2.2 Classical Holophrasm

Now that we have seen the structure of proof trees, we will explain the search used by Holophrasm [9] through this trees.

Algorithm. The algorithm visits the root using $VisitNodeOR$ as long as this root has not been proven or the number of his visits has not reached a certain threshold. The value of the root is, as for all OR nodes, a probability calculated by the payout neural network modeling the chance of proving the OR node with the hypotheses it contains. When visiting an OR node, the objective is then to visit an AND node (proposition), which has the highest chance of being provable. Note that for an AND node, its probability is given by the prediction neural network and the substitution is given by the generative network. Furthermore, when an AND node is created, all its children are also created and visited once. Initial values are given by two different networks, depending on the nature of the node. The payout network takes an OR node as an argument and outputs the probability that this node is provable with the assumptions it contains. The prediction network gives a softmax on the set of propositions applicable to an OR node. Another feature we would like to detail is the interface adding an AND node to an OR node. Holophrasm uses a heap in which all potential candidates are stored. A visit to an OR node may not add an AND node, if the candidate is not valid. An OR node will be considered disproven if all its children are disproof, or if it has been visited enough times and the heap is empty. An AND node is disproven if one of its childs is disproven. Disproven AND node are cut from the tree.

Algorithm 1: Research Function of Holophrasm

```
1  Function Holophrasm(node, maxpasses):
2      passes = 0;
3      while (not node.proven) or passes<maxpasses do
4          VisitNodeOR(node);
5          UpdateProven(node, "OR");
6          UpdateValueOR(node);
7      end
```

Algorithm 2: Visit of an "AND" node by Holophrasm

---

1 **Function** VisitNodeAND(*node*):
2   **if** *len(node.children)* > 0 **then**
3   VisiteNodeOR(node.children[argmin([child.value/child.visit for child in node.children])]);
4   **end**
5   UpdateProven(node, "AND");
6   UpdateValueAND(node);

Algorithm 3: Update of an "AND" node by Holophrasm

---

1 **Function** UpdateValueAND(*node*):
2   **if** *len(node.children)* > 0 **then**
3   badchild = node.children[argmin([child.value/child.visit for child in node.children])];
4   node.visit = badchild.visit;
5   node.value = badchild.value;
6   **end**
7 **return** 0;

Algorithm 4: Visit of an "OR" node by Holophrasm

---

1 **Fonction** VisitNodeOR(*node*):
2   **if** *node.heap is empty* **then**
3   Use the Holophrasm Interface to fill node.heap of all compatible proposition;
4   **end**
5   **if** *node.visit/6 + 0.01 > len(self.children) + node.childlessvisit* **then**
6   Try to add an "AND" node from the heap, if it fail node.childlessvisit+=1;
7   **return** 0;
8   **end**
9   **if** *len(node.children)* > 0 **then**
10   value = 0;
11   nextchild = None;
12   **for** *child in node.children* **do**
13   **if** *valuation-function(node.visit, child)* > *value* **then**
14   value = valuation-function(node.visit, child);
15   nextchild = child;
16   **end**
17   VisiteNodeAND(nextchild);
18   **end**
19   **end**
20   UpdateProven(node, "OR");
21   UpdateValueOR(node);

Algorithm 5: Update of an "OR" node by Holophrasm

---

1 Function UpdateValueAND(*node*):
2 | if *len(node.children)* > 0 **then**
3 | | node.visit = sum(child.visit for child in node.children) + node.childlessvisit + 1;
4 | | node.value = node.networkpayout + sum(child.value for child in node.children);
5 | **end**

Algorithm 6: Update if a node is proven

---

1 Function UpdateProven(*node, type*):
2 | if *type == "OR"* **then**
3 | | if *len(node.children) == 0 and node.visit> 1 and node.heap is empty* **then**
4 | | | node.dead = True ;
5 | | **end**
6 | | **for** *child in node.children* **do**
7 | | | if *child.proven* **then**
8 | | | | node.proven = True;
9 | | | **end**
10 | | **end**
11 | **end**
12 | if *type == "AND"* **then**
13 | | node.proven = True;
14 | | **for** *child in node.children* **do**
15 | | | if *not child.proven* **then**
16 | | | | node.proven = False;
17 | | | | break;
18 | | | **end**
19 | | **end**
20 | | if *any ([child.dead for child in node.children])* **then**
21 | | | CutNodeFromTree(node);
22 | | | return 0;
23 | | **end**
24 | **end**

Algorithm 7: Evaluation Function for guiding the exploration in Holophrasm

---

1 Function valuation-function(*fathervisit, nodeAND*):
2 | return $\frac{nodeAND.value}{nodeAND.visit+1} + 0.5 * \frac{nodeAND.probabilitynet}{1+nodeAND.visit} + \sqrt{\frac{\log(fathervisit)}{1+nodeAND.visit}}$,

---

Results. On the first 200 theorems of the Holophrasm's Test set and with the parameter 10 for the BeamSearch used by the generative network, we obtain Fig. 1.

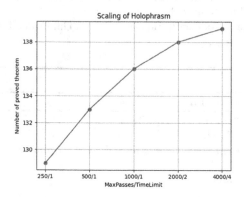

Fig. 1. Results Holophrasm

We are not going to make a detailed analysis of the results, which we will use mainly to compare with the other algorithms we are going to test.

3   Classical Tree Search Algorithm for Metamath Theorem Proving

In this section we will take a look at a number of well-known algorithms applied in this context by changing the Holophrasm search. We will start by analysing Minimax and its results, then we will study more selective algorithms such as PUCT.

3.1   Minimax

Now that we have tested the search algorithm provided with the Holophrasm interface, we are going to test a more traditional search: Minimax. The aim is to compare a more conventional search algorithm with the Holophrasm's results, and highlight the importance of progressive widening (the expansion of OR node's child).

Algorithm. We launch a Minimax search from the root by setting the depth. However, the problem is that an OR node can have an infinite number of children. To overcome this problem, we limit the number of possible children to a fixed breadth. The children of the OR node are chosen according to their probability given by the prediction neural network. The depth is reduced by 1 each time we go from an AND node to an OR node, and the algorithm ends on OR nodes.

With regard to node initialization, when an AND node is created, all its OR child nodes are created and checked if they are an initial hypothesis of the problem. The point of testing several breadths is to evaluate the efficiency of the network in finding the next proposition to apply to an OR node.

Results. Our test set consists of the first 200 Holophrasm theorems. The test conditions are 1 pass in the maximum tree and a parameter of 10 for the Beam-Search used in the Holophrasm interface.

	depth = 2	depth = 3
breadth = 2	78/200	79/200
breadth = 3	94/200	95/200
breadth = 4	97/200	98/200
breadth = 5	103/200	108/200
breadth = 7	112/200	113/200
breadth = 9	117/200	.../200

From the results, we can see that the networks are often wrong, and that we need to go to wider breadths to get better results. This underlines the importance of the progressive widening used in Holophrasm.

3.2   PUCT

We now test other conventional algorithms such as PUCT [7], Product Propagation [6] and Proof Number Search [1]. These tree search methods are more recent than Minimax and they allows one to find a solution without exploring the whole tree. Some of these approaches seams to not use networks, but in fact they are used implicitly by the interface when it attributes an And node to an Or node.

Algorithm. The idea behind PUCT is to supervise the search when choosing the next AND node to visit. This algorithm is inspired by the Bandit literature. To implement PUCT from Holophrasm's research, we will just change the Holophrasm bandit by changing the valuation-function.

Results. First, we will determine the best constant to use in PUCT. With the same BeamSearch setting as before and on the parameters (1000 passes, 1 min), we obtain:

Algorithm 8: Evaluation Function for PUCT

Result: PUCT

1 **Function** valuation-function(*fathervisit*, *nodeAND*):

2 | Return $\frac{nodeAND.value}{nodeAND.visit} + C * \frac{nodeAND.prediction\_net}{1+nodeAND.visit} * \sqrt{fathervisit}$;

C	0.1	0.2	0.3	0.4	0.5
Results	136/200	136/200	136/200	136/200	135/200

We will therefore test the Scaling of PUCT with $C = 0.2$, the results are described in Fig. 2.

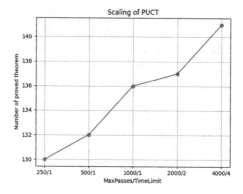

Fig. 2. Results PUCT for C=0.2

In comparison with Holophrasm, the results are quite similar except for the parameters (4000,4). So, with less time constraint, PUCT does better than Holophrasm. It is difficult to explain this behavior further. We will see below that there is a kind of ceiling around 140, whatever the algorithm used.

3.3 Product Propagation

Since PUCT is an algorithm that affects exploration, we will now look at Product Propagation (PP) [6], which mainly changes the value and update of nodes.

Algorithm. The idea is to change the visit and the values attributed to the nodes during the search. The value is seen as a probability and the children are assumed to be independent. Thus the value of an AND node is the product of the values of its children and the value of an OR node is calculated using the same principle but with the additional probability. In an OR node we explore the best valued child, and in an AND node the worst. The value of Leaf can be initialized with 1 or by the payout network given by Holophrasm, but in our case we used the payout network for better results.

Results. With the same settings as before we obtain Fig. 3. The results show, in this context, that PP is more efficient than PUCT on parameters where the number of passes is limiting. However, for the parameter (4000,4), we achieve the same result as PUCT. PP seems to be more efficient, proving 138 propositions in just 1 min and 500 passages maximum.

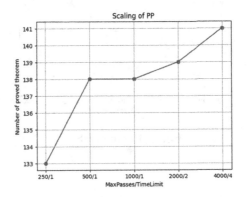

Fig. 3. Results Product Propagation

3.4   Proof Number Search

Similar to PP, Proof Number Search (PNS) [1] is based on node values and updates. However, this time there are two values per node to take into account.

Algorithm. The idea is to count the number of leaves left to explore either to prove the node or to disprove it. During the node initialization, a non proven node is initialized with a Proof Number and a Disproof Number. In the original algorithm, for a non-proven or proven node, the PN and DPN are initialized to 1. For a proven node, the PN is set to 0 and the DPN to infinity, and vice versa for a proven node. Then during the visit, in an OR node we choose the AND node with the lowest DPN and in an AND node we choose the OR node with the lowest PN. One might think that neural networks are not used in this algorithm, but they are used indirectly in AND node assignment.

Results. With the same settings as before, we obtain Fig. 4. The results are weaker than Product Propagation and all its variants (which we will see later), but in this test we are not using the value given by the "payout" neural network. We tried to improve PNS, but we could not find any approach that improve the results presented above, either by initializing PN and DPN with the networks, or by using a PUCT in the search, so we haven't included these approaches in this article.

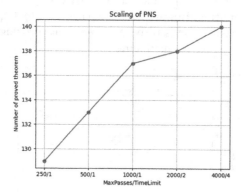

Fig. 4. Results PNS

3.5 HyperTree Proof Search

We are now going to test an algorithm which is different in his approach from the previous ones. This algorithm try to obtain a broader perspective by selecting wider sub-tree instead of doing descent.

Algorithm. The goal is to draw inspiration from the search algorithm presented in [4], in order to compare the results of different approaches in this context. However, an adaptation is necessary because the interfaces used are different. The idea is to select a sub-tree of the proof tree, expand the leaves, then update only the values of the AND nodes of this sub-tree using Product Propagation. Transposed to our interface, this is like using PUCT to select AND nodes, but once in an AND node, expand all its OR children.

Results. With 10 for the interface BeamSearch, we obtain the results described in Fig. 5.

The results, in this context, are inferior to those of the previous algorithm. Extending all OR nodes that are children of an AND node is time-consuming. This explains the jump in performance from 2 to 4 min. This may be the algorithm with the worst results, but when you look at the evolution of the curve, it is the one that improves the most as a function of time.

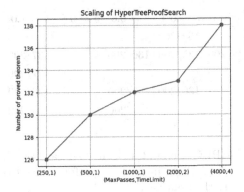

Fig. 5. Results HyperTree Proof Search

4 New Tree Search Algorithm for Metamath Theorem Proving

Now that we have seen the more traditional approaches, we are going to use them as an inspiration to create new ones.

4.1 Production Propagation Combined with PUCT

Algorithm. Given the performance of Product Propagation and PUCT, the idea is to combine the two approaches. In fact, these two algorithms should work well together, since Product Propagation does not rely on a bandit's part. We'll present the two combinations we've tested. For the first algorithm, the Product Propagation value is used, along with the PUCT visit and bandit. In the case of the second approach, we change the PUCT bandit (see algorithm 9). The aim was to keep the first Holophrasm bandit approach, but combine it with PUCT. This is useful especially when the policy given by the payout network is greatly underestimating a son. In this case the part coming from UCB compensates.

Algorithm 9: Modified Evaluation function for PUCT

---

1 **Function** modified-valuation-function(*fathervisit, nodeAND*):

2   Return $nodeAND.value$

   $+a * node.probabilitynet * \frac{\sqrt{fathervisit}}{nodeAND.visit}$

   $+b * \sqrt{\frac{\log(fathervisit)}{nodeAND.visit}};$

Results. With 10 for the interface BeamSearch and the modified validation function we obtain:

	a = 0.1	a = 0.4	a = 0.6	a = 0.8	a = 1.0	a = 1.1
b = 0.1		134/200				
b = 0.4	133/200	135/200	136/200		137/200	
b = 0.5				138/200	137/200	136/200
b = 0.6		130/200	136/200		136/200	
b = 1.0			134/200		135/200	

In the case of the PUCT bandit (Fig. 6), we can see that combining the two approaches results in a slight increase in performance. However, we are still unable to break the 141-theorem barrier.

There are two interesting points to note about the modified PUCT bandit (Fig. 7). Firstly, the algorithm performs particularly well with parameters (4000,4), where it proves 144 theorems (and thus breaks the 141-theorem barrier). It is therefore the best on these criteria. Note that, this bandit exploration term has been used with other search algorithms, including Holophrasm, but its results were not convincing. The final point is that if we remove the principle of node unprovability (see the UpdateProven 24 code line 3–4 and 20–22), this algorithm is the one with the best results.

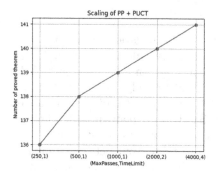

Fig. 6. Results PP + PUCT c=0.2

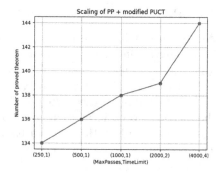

Fig. 7. Results PP + PUCT with modified validation function a = 0.8 b = 0.5

5   Conclusion

We studied different algorithms applied to theorem proving in Metamath. Minimax algorithm highlighted the importance of progressive widening. Algorithms such as PUCT, PNS or PP obtained better results than the search used by Holophrasm. Finally, we propose a new algorithm by combining the idea of PUCT and PP and obtain better results.

Our initial goal was to improve the search used by the Holophrasm interface. The different algorithms we proposed, as well as the different solutions we provided, enabled this improvement. Although the improvement is slight (see Fig. 8), our different approaches seem more promising in terms of parameter-dependent changes in performance. In future work we plan to test this hypothesis by applying the algorithms on the whole dataset.

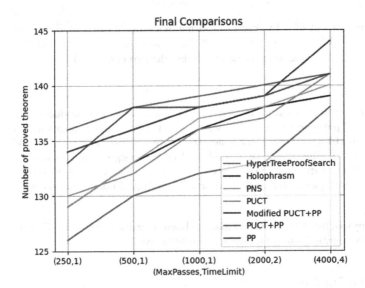

Fig. 8. Final Comparisons

References

1. Allis, L.V., van der Meulen, M., van den Henrik, H.: Proof-number search 1 (2003)
2. Browne, C., et al.: A survey of Monte Carlo tree search methods. IEEE Trans. Comput. Intell. AI Games **4**(1), 1–43 (2012). https://doi.org/10.1109/TCIAIG.2012.2186810
3. Fawzi, A., et al.: Discovering faster matrix multiplication algorithms with reinforcement learning. Nature **610**(7930), 47–53 (2022)
4. Lample, G., et al.: Hypertree proof search for neural theorem proving. In: Neural Information Processing Systems (NeurIPS) (2022)
5. Megill, N.D.: Metamath: A Computer Language for Mathematical Proofs. Lulu Press, Morrisville, North Carolina (2019)
6. Saffidine, A., Cazenave, T.: Developments on product propagation. In: van den Herik, H.J., Iida, H., Plaat, A. (eds.) CG 2013. LNCS, vol. 8427, pp. 100–109. Springer, Cham (2014). https://doi.org/10.1007/978-3-319-09165-5_9
7. Silver, D., et al.: Mastering the game of go with deep neural networks and tree search. Nature **529**, 484–489 (2016)
8. Silver, D., et al.: A general reinforcement learning algorithm that masters chess, shogi, and go through self-play. Science **362**(6419), 1140–1144 (2018)
9. Whalen, D.: Holophrasm: a neural automated theorem prover for higher-order logic. arXiv preprint arXiv:1608.02644 (2016)
10. Wu, D.J.: Accelerating self-play learning in go. arXiv preprint arXiv:1902.10565 (2019)

Slitherlink Art

Cameron Browne[✉]

Department of Advanced Computing Sciences, Maastricht University,
Paul-Henri Spaaklaan 1, 6229 EN Maastricht, The Netherlands
cameron.browne@maastrichtuniversity.nl

Abstract. Slitherlink is a pure deduction puzzle that produces simple
closed paths in a square grid when solved. This paper presents a simple
method for automatically generating Slitherlink challenges whose solutions describe artistic shapes selected by the designer, and its successful
implementation in a Java program. The process is assessed through several examples in terms of their artistic merit and quality as puzzles.

Keywords: Slitherlink · Japanese Logic Puzzles · Deductive Search ·
Procedural Content Generation · Subtractive Puzzle Design

1 Introduction

Slitherlink is a popular pure deduction puzzle in the class of *Japanese logic
puzzles* [13] in which players must draw a simple closed path between the points
in a grid that visits each hint cell on the specified number of sides. Figure 1 shows
a small example and its solution (from [15]). Each challenge should provide a
single unique solution, like all well-designed examples of this class of puzzle [3].

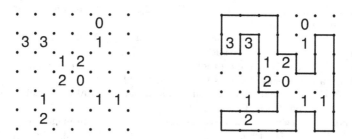

Fig. 1. A Slitherlink challenge (left) and its solution (right) from [15].

The hints of a challenge are typically arranged symmetrically (see Fig. 1,
left), but the solution path thus defined is usually arbitrarily shaped and bears
little resemblance to the hint pattern. A well-designed hint set will give as few
clues as possible to the solution path shape, in order to challenge the player.

© The Author(s), under exclusive license to Springer Nature Switzerland AG 2024
M. Hartisch et al. (Eds.): ACG 2023, LNCS 14528, pp. 158–170, 2024.
https://doi.org/10.1007/978-3-031-54968-7_14

Occasionally, some Slitherlink challenges are deliberately designed so that their solution paths form recognisable shapes, e.g. the designer's initials, the outline of a christmas tree for a Christmas special, and so on. This paper extends this idea to explore how artistic solution paths based on given pictures may be embedded in standard Slitherlink challenges, while keeping them deducible and interesting. The aim is to surprise and delight players[1] as they produce works of art through the solution process. The two main research questions are:

RQ1: Can Slitherlink challenges be designed to produce artistic solution paths while remaining interesting for human players to solve?

RQ2: Can the design of such artistic Slitherlink challenges be automated?

2 Prior Work

Fig. 2. "Mario" Picture Slitherlink from [10].

Opp [10] has previously devised "Picture Slitherlink" examples that reveal pixelated images in the solution path, such as the challenge shown in Fig. 2 (left) which produces an image of the Mario character (shrunken and filled on the right) when solved. Opp provides three other examples [10], along with a program that allows users to manually load, view, edit, solve and save compatible Slitherlink challenges without automating any of these tasks. This work shows that the task is feasible, but leaves room for improvement:

---

[1] I use the term "players" rather than "solvers" throughout, in order to avoid confusion with automated problem solvers.

- The resulting figure is not very clear and is obscured by the background fill.
- The hint set is asymmetrical, unevenly spread, and reveals the solution shape to some extent.
- The challenge is too hard to interest human players (see below).
- The design process is not automated.

2.1 Deductive Search

Deductive Search (DS) is a constraint-based search method for solving deduction puzzles phrased as constraint satisfaction problems, which emulates the processing limits experienced by human players [4]. It involves three basic operations:

1. *Simplification:* Eliminate values that contradict known constraints.
2. *Shaving:* Eliminate values whose realisation would produce a contradiction.
3. *Agreement:* Resolve values that agree regardless of remaining resolutions.

DS works by first applying all obvious (Level 0) simplifications to the puzzle's initial state. It then applies a series of Level 1 hypotheses by trying each remaining value for each unresolved variable, applying the resulting simplifications, eliminating those values whose realisation would cause a contradiction (Level 1 shaving) and resolving other variables that yield the same value for every hypothesis (Level 1 agreement). If this does not resolve all variables then the process is repeated recursively with Level 2 deductions applied at each step of the Level 1 deduction process.

A puzzle challenge is deemed *deducible* if this process yields a solution, which is then guaranteed to be unique [3]. Importantly, the process does not recognise solutions stumbled upon during higher level simplifications; it only recognises *solutions provided by pure deduction*.

Recursive structures (such as Level 2 deductions) are hard for humans to model mentally [5] and evidence from psycholinguistics suggests a mental limit of two recursive levels of embedding in natural language processing [8]. This cognitive limit also appears to apply to puzzle solving, as a study of DS applied to a range of published puzzles of different types (including Slitherlink) revealed that most cases tested could be solved with Level 1 deductions, with some more difficult cases requiring a small number of Level 2 deductions [4]. Anything beyond this limit would be exceedingly difficult for human players.

2.2 Difficulty

Pelánek [11] describes a method for estimating the difficulty of Sudoku challenges for players based on: (1) the complexity of individual steps, and (2) the structure of dependency among those steps. DS provides a similar approach for estimating the difficulty of deduction puzzles based on the number and type of deductive steps applied during the solution process – agreement requires more mental processing than shaving – and the levels at which they occur [4].

Slitherlink is inherently difficult as it requires considerable visualisation, spatial reasoning and a combination of local and global planning. For example, the small challenge shown in Fig. 1 is described as "moderately difficult" [15] although it is solvable with a few Level 1 deductions. See Sect. 2.1 for an explanation of deduction Levels 1, 2 and 3.

With this in mind, DS reveals the "Mario" Picture Slitherlink shown in Fig. 2 to be too difficult for human players. DS yielded a higher difficulty estimate (2.829) for this challenge after 47.5 s, much higher than any other Japanese logic puzzle tested with DS [4]. This challenge is deducible but requires many Level 2 deductions including a daunting 48 Level 2 agreements (see Table 1) which would be exceedingly difficult for even the most mentally gifted and experienced player. Even if this was achievable, it would make this challenge an extremely taxing task rather than an enjoyable one.

Table 1. Deductive Search results for the "Mario" Picture Slitherlink.

Deduction	Simplifications	Shavings	Agreements
Level 0	244	—	—
Level 1	531	113	58
Level 2	442	17	48

2.3   Interestingness

The difficulty of a given puzzle challenge is closely related to how interesting the player will find it; a trivial challenge will not engage the player while one that is too difficult will be frustrating rather than enjoyable. Nobuhiko Kanamoto, Chief Editor of Nikoli,[2] the world's foremost publisher of Japanese logic puzzles, observes that the level of difficulty should also be consistent throughout the solution process, and presents a poor Sudoku challenge whose solution involves a difficult deduction followed by a series of trivial deductions until its solution [7]. Such inconsistency is a characteristic of low quality computer-generated challenges, and is one reason why Nikoli only publish handcrafted challenges for their puzzles. Higashida [6] further observes that automated mass production of low quality puzzle challenges is a threat to puzzle design as an art form.

Some superficial aspects that increase the perceived quality of puzzle challenges for players are easy to achieve, such as generating hint sets that are symmetrically or artistically arranged, provided that this does not introduce repetitions in solution strategies [1]. An interesting puzzle will demonstrate *dependency* by offering a series of mini-puzzles whose solution yields information that opens up further mini-puzzles, and so on, until the solution is achieved [2]. Further, a well designed puzzle will heighten the drama with cycles of tension in which

---

[2] https://www.nikoli.co.jp/en/puzzles/.

information dries up to give crisis points whose resolution suddenly opens up new parts of the challenge to resolve [4]. To maintain lasting interest, a puzzle should have the potential for increasingly subtle and effective strategies for players to learn [9] and well-designed challenges should incorporate as wide a spread of these strategies in their solutions process as possible, although this aspect is beyond the scope of this study.

3   Approach

The basic approach for automating the design of Slitherlink art is shown pictorially in Fig. 3. Starting with a provided source image (left), the image is undersampled to a desired grid size[3] and grid cells corresponding to foreground pixels in the image are turned "on" to define a *solution set* of connected cells. Each cell is then assigned a number indicating the number of its edges on the solution set boundary. A subset of these numbers is then selected to be shown as the *hint set* for the challenge.

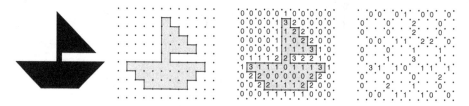

Fig. 3. The source image (left) is pixelated to a solution set, boundary edges counted per cell, then a subset of these is selected to be shown as the challenge hint set (right).

3.1   Boundary Repair

One caveat is that the solution set must constitute a single 4-connected (i.e. orthogonally connected) set of cells if the resulting path is to be single, simple (i.e. not intersecting) and closed. These conditions are a requirement of the game and ensure that the solution path has a distinct inside and outside as per the Jordan Curve Theorem [12], which allows some helpful solution strategies [4].

Figure 4 shows the steps to guarantee these conditions. Once the initial solution set is obtained (Fig. 4, left), a check is made for any 2×2 patterns with two on-pixels and two off-pixels at opposed diagonals (such as the ears near the top or the leg joints near the bottom of the figure). Such cases are resolved by flipping the value of the cell within each 2×2 pattern that expands the set as much as possible while maintaining overall 4-connectivity.

---

[3] The grid size is defined by a *cellSize* parameter as a percentage of the maximum image dimension, i.e. *cellSize*=5% gives a grid with 20 cells horizontally or vertically.

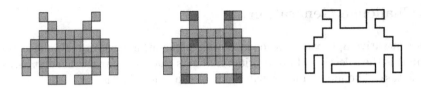

Fig. 4. The solution set is augmented to yield a simple closed boundary.

A check is then made on the number of boundary paths around the solution set, and all except the largest such path discarded (such as the two eye holes in the figure). This step is applied separately to both on-pixel and off-pixel boundaries. The resulting solution set boundary is then guaranteed to be a single, simple, closed path (Fig. 4, right).

3.2   ASCII Art Sources

Since pixelating the source image to the challenge grid inevitably degrades the figure shape, and boundary repair can degrade it even further, the puzzle designer can instead upload solution sets as plain text "ASCII art" files. The designer can manually edit the exact solution set they want (e.g. Figure 5), although this process is tedious and time consuming compared to loading solution images and automatically pixelating them.

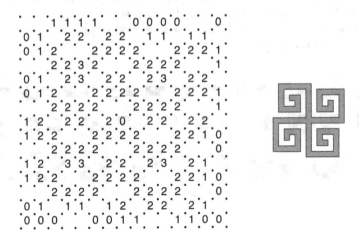

Fig. 5. A manually prepared solution set (right) allows precise placement of on-pixels.

4   Challenge Generation

This generative approach outlined above is straightforward and simple to program; it is the selection of appropriate hint sets that is by far the hardest aspect of this task. The ideal hint set will define a challenge that is:

- *Deducible*: Can be solved purely by deduction.
- *Minimal*: Most or all hints shown should be necessary for solution. Hint sets should typically cover less than 50% of the grid.
- *Achievable*: At most Level 1 deductions.
- *Challenging*: At least some Level 1 deductions.
- *Dependent*: Deductions should reveal information that leads to further deductions [2].
- *Tense*: There should not be too many deduction points available at once [4] [2].
- *Evenly Spread*: Hints should be evenly spread across the grid.
- *Artistic*: The hint set should be attractive, symmetric and possibly bear some relevance to the hidden solution figure.

The following hint types are implemented in the Slitherlink Art program available at: http://www.cambolbro.com/research/Slitherlink_Art_1.0.jar

4.1   Repeated Hint Patterns

The simplest hint pattern type implemented allows the user to select a sub-pattern from a predefined library and apply that sub-pattern repeatedly across the grid modulus its width and height. Figure 6 shows the available library of sub-patterns and Fig. 7 shows three of these applied to produce the boat shape shown above in Fig. 3 (right).

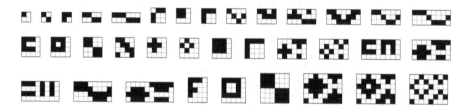

Fig. 6. The library of predefined hint patterns that may be repeated across the grid.

Repeated hint patterns are fast to generate and test, although sparser sub-patterns with less than 50% coverage often fail to produce a deducible result. Hence an "Invert Hints" option is provided so that greater coverage can be readily achieved to more likely produce a deducible result.

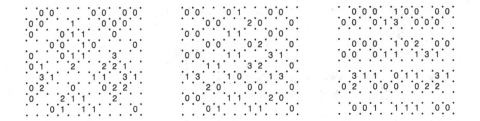

Fig. 7. Repeated hint patterns that produce the same solution as Fig. 3 (right).

4.2 Radial Hint Patterns

The radial hint pattern is obtained by visiting each cell in the grid and turning on those whose distance to the grid centre rounded to the nearest integer is even. This hint pattern rarely produces a deducible result and it is typically useful to automatically iterate over a range of cell sizes to find any that produce a good result. For example, The 15×15 challenge shown in the Fig. 8 was the largest grid size found to produce a deducible hint set for the pi figure.

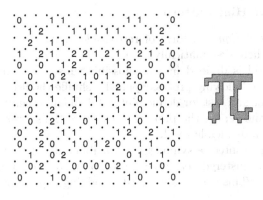

Fig. 8. Concentric circles producing the symbol for pi.

4.3 Spiral Hint Patterns

The spiral hint pattern is obtained by starting at the grid's central cell and following a path that travels for $n = 1$ steps in a given direction then iteratively turns 90° clockwise and increments n until the grid is covered, turning on cells that are visited along the way.

Spiral hint sets tend to be more deducible than radial ones, probably due to their longer consecutive lines of orthogonally adjacent hints, but are wasteful in the number of superfluous 0 hints that give away the solution shape (see Fig. 9).

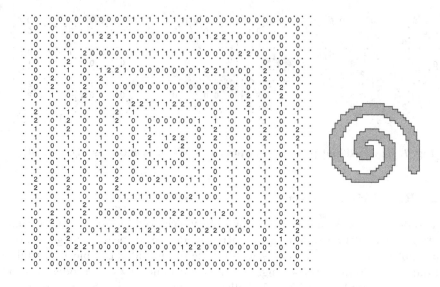

Fig. 9. A square spiral hint set producing a round spiral figure.

4.4 Subtractive Hint Patterns

Subtractive puzzle design is a form of retrograde analysis which starts with a chosen solution then systematically removes hints while the solution is still deducible. The grid is divided into all possible subsets of four symmetrically placed cells and the following process applied – shuffle the list then successively remove each subset and test for deducibility; if deducible then remove that subset from the list and repeat the process else restore the subset – until no more removals produce a deducible result.

This process guarantees a symmetric deducible hint set that is close to minimal, but typically clustered near the solution path to give an uneven spread of hints (Fig. 10, left). This can be addressed by adding superfluous 0 hints evenly

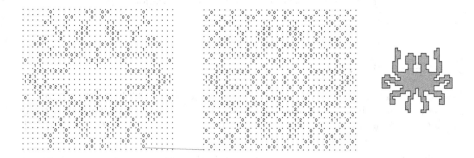

Fig. 10. Subtractive hint removal (left) with additive area filling (right).

across empty regions, e.g. using a standard "chessboard" distance metric, to give a more homogenous spread and obscure the solution shape (Fig. 10, right). Subtractive design is time-consuming and takes at least several minutes per case.

4.5   Pictorial Hint Patterns

The artistic aspect can be emphasised by making the challenge hint set pictorial as well. This is easily achieved by loading a source image (which is pixelated to the grid size) or ASCII art text file to define the on-cells in the hint set. Figure 11 shows two such picture-to-picture examples. Note that there are no requirements on the connectivity of the hint set, so the boundary repair process described above does not need to be applied to it, just to the solution set.

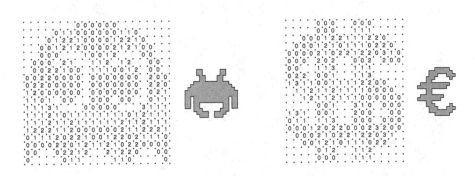

Fig. 11. Picture-to-picture puzzles: Pacman to space invader and dollar to euro.

5   Summary

Table 2 provides a summary of typical performance of each of the main hint pattern types in producing Slitherlink Art. The categories are subjective but reflect the typical performance observed. If artistic merit is the priority, then repeating hint patterns proved effective in quickly producing attractive hint sets that yielded deducible challenges of non-trivial difficulty. If puzzle quality is

Table 2. Typical performance of each hint pattern method in practice.

Hint Type	Speed	Deducible	Artistic	Minimal	Difficult	Interesting
Repeating	Fast ($\approx$1 s)	Often	Often	No	Sometimes	Sometimes
Radial	Fast ($\approx$1 s)	Rarely	Often	No	Rarely	Sometimes
Spiral	Fast ($\approx$1 s)	Often	Often	No	Rarely	Sometimes
Subtractive	Slow (mins)	Always	Sometimes	Yes*	Usually	Usually
From Picture	Fast ($\approx$1 s)	Rarely	Always	No	Rarely	Rarely

Fig. 12. A high resolution dragon with repeated inverse wave hint pattern.

the priority, then the subtractive approach proved most successful, producing symmetric and minimal hint sets (* unless area filling is applied) that yield difficult and interesting challenges but can take considerable time to generate.

The automated process produces artistic results easily, e.g. the dragon challenge shown in Fig. 12 was generated in less than a second using a hint pattern from the predefined library (Fig. 6, top right). The process emphasises puzzle as artwork and presents solution shapes clearly without background clutter, providing no guarantees on the interestingness of the resulting challenges. However, feedback from players suggests that the novelty of artistic shapes emerging from the solution process increases their enjoyment of Slitherlink Art challenges.

6   Conclusion

In terms of Research Question 1, it proved relatively straightforward to produce Slitherlink challenges with artistic solution paths with the potential to interest human players. In terms of Research Question 2, the process can indeed be readily automated. The subtractive design approach proved most effective at generating minimal, symmetric, difficult and interesting Slitherlink Art challenges but was also the most time-intensive. The repeated pattern approach proved most effective at quickly generating symmetric challenges without guarantees on their potential to interest human players. The picture-to-picture variation takes Slitherlink Art to another artistic level but at the expense of puzzle quality.

Future work might focus on fine-tuning hint sets to produce truly minimal hint sets and better guarantees on the interestingness of challenges (e.g. based on dependencies and tension during the solution process). It might also be fruitful to investigate search-based *procedural content generation* (PCG) methods, such as evolutionary approaches, in addition to the constructive approaches described here [14]. The examples presented here demonstrate the feasibility of the idea but further work is needed to guarantee the quality of the resulting challenges as actual puzzles as well as works of art.

Acknowledgements. Thanks for Stephen Tavener for pointing out the Picture Link examples [10] and Néstor Romeral Andrés for suggesting the picture-to-picture variation. This work was partly funded by the European Research Council (ERC CoG #771292).

The Java program implemented for this study is available at: http://www.cambolbro.com/research/Slitherlink_Art_1.0.jar

References

1. Browne, C.: Metrics for better puzzles. In: El-Nasr, M.S., Drachen, A., Canossa, A., Isbister, K. (eds.) Game Analytics: Maximizing the Value of Player Data, pp. 769–800. Springer, Berlin (2013). https://doi.org/10.1007/978-1-4471-4769-5_34
2. Browne, C.: The nature of puzzles. Game Puzzle Des. **1**(1), 23–34 (2015)
3. Browne, C.: Uniqueness in logic puzzles. Game Puzzle Des. **1**(1), 35–37 (2015)

4. Browne, C.: Deductive search for logic puzzles. In: 2013 IEEE Conference on Computational Intelligence and Games (CIG 2013), pp. 359–366. Niagara Falls (2013)

5. Corballis, M.: The uniqueness of human recursive thinking. Am. Sci. **95**, 240–248 (2007)

6. Higashida, H.: Machine-made puzzles and hand-made puzzles. In: Advances in Information and Communication Technology (IFIP AICT 333), vol. 333, pp. 214–222. Brisbane (2010)

7. Kanamoto, N.: A well-made sudoku is a pleasure to solve. https://www.nikoli.co.jp/en/puzzles/sudoku/why_hand_made/ (2001–2021)

8. Karlsson, F.: Syntactic recursion and iteration. In: van der Hulst, H. (ed.) Recursion and Human Language, pp. 43–67. Mouton de Gruyter, Berlin (2010)

9. Lantz, F., Isaksen, A., Jaffe, A., Nealen, A., Togelius, J.: Depth in strategic games. In: AAAI 2017 Workshop on What's Next for AI? AAAI Press, San Francisco (2017)

10. Opp, J.R.: Picture Slitherlinks!, Caravel Forum (2015). https://forum.caravelgames.com/viewtopic.php?TopicID=39535

11. Pelánek, R.: Difficulty Rating of Sudoku Puzzles: An Overview and Evaluation. Technical report, Faculty of Informatics, Masaryk University Brno, Czech Republic (2014)

12. Spanier, E.: Algebraic Topology. McGraw Hill, New York (1966)

13. Times, T.: Japanese Logic Puzzles: Hashi, Hitori, Mosaic and Slitherlink. Harper Collins, London (2006)

14. Togelius, J., Yannakakis, G.N., Stanley, K.O., Browne, C.: Search-based procedural content generation: a taxonomy and survey. IEEE Trans. Computa. Intell. AI Games **3**(3), 172–186 (2011)

15. Wikipedia: Slitherlink, 28 August 2023. https://en.wikipedia.org/wiki/Slitherlink

Author Index

© The Editor(s) (if applicable) and The Author(s), under exclusive license
to Springer Nature Switzerland AG 2024
M. Hartisch et al. (Eds.): ACG 2023, LNCS 14528, p. 171, 2024.
https://doi.org/10.1007/978-3-031-54968-7

Printed in the United States
by Baker & Taylor Publisher Services

Printed in the United States
by Baker & Taylor Publisher Services